Deepak Chopra, M.D., is the author of mo... ...ixty-five books, including numerous *New York Times* bestsellers. His medical training is in internal medicine and endocrinology, and he is a Fellow of the American College of Physicians, a member of the American Association of Clinical Endocrinologists, Adjunct Professor of Executive Programs at the Kellogg School of Management at Northwestern University, and Distinguished Executive Scholar at the Columbia Business School, Columbia University. Since 1997 he has participated annually as a lecturer at the Update in Internal Medicine event sponsored by Harvard Medical School, Department of Continuing Education, and the Department of Medicine, Beth Israel Deaconess Medical Center. Deepakchopra.com

Rudolph E. Tanzi, Ph.D., is the Joseph P. and Rose F. Kennedy Professor of Neurology at Harvard University, and Director of the Genetics and Aging Research Unit at Massachusetts General Hospital (MGH). Dr. Tanzi has been investigating the genetics of neurological disease since the 1980s, when he participated in the first study using genetic markers to find a disease gene (Huntington's disease). Dr. Tanzi isolated the first Alzheimer's disease gene and discovered several others; he now heads the Alzheimer's Genome Project. He is currently developing promising new therapies for Alzheimer's disease. Dr Tanzi serves on dozens of editorial and scientific advisory boards, and chairs the Cure Alzheimer's Fund Research Consortium. He has received numerous awards, including the two highest awards for Alzheimer's disease research: the Metropolitan Life Award and the Potamkin Prize. Dr. Tanzi has coauthored more than four hundred scientific research articles and book chapters. He also coauthored the book *Decoding Darkness: The Search for the Genetic Causes of Alzheimer's Disease.*

Also By Rudolph E. Tanzi

Decoding Darkness (with coauthor Ann B. Parson)

Also By Deepak Chopra

Creating Health

Return of the Rishi

Quantum Healing

Unconditional Life

Journey into Healing

Creating Affluence

Perfect Weight

Restful Sleep

The Seven Spiritual Laws of Success

The Return of Merlin

Boundless Energy

Perfect Digestion

The Way of the Wizard

Overcoming Addictions

Raid on the Inarticulate

The Path to Love

The Seven Spiritual Laws for Parents

The Love Poems of Rumi
(edited by Deepak Chopra; translated
by Deepak Chopra and Fereydoun Kia)

Healing the Heart

Jesus

Reinventing the Body, Resurrecting the Soul

The Ultimate Happiness Prescription

Muhammad

The Soul of Leadership

The Seven Spiritual Laws of Superheroes
(with coauthor Gotham Chopra)

*Consciousness in the Universe: Quantum Physics,
Evolution, Brain and Mind* (with Stuart Hameroff and
Sir Roger Penrose)

Is God An Illusion? (with co-author
Leonard Mlodinow)

Self Power

God: A Story of Revelation

For Children

On My Way to a Happy Life (with Kristina Tracy,
illustrated by Rosemary Woods)

You with the Stars in Your Eyes
(illustrated by Dave Zaboski)

SUPER
BRAIN

UNLEASHING THE EXPLOSIVE POWER OF
YOUR MIND TO MAXIMIZE HEALTH,
HAPPINESS, AND SPIRITUAL WELL-BEING

DEEPAK CHOPRA

AND

RUDOLPH E. TANZI

RIDER

LONDON · SYDNEY · AUCKLAND · JOHANNESBURG

1 3 5 7 9 10 8 6 4 2

Published in 2012 by Rider, an imprint of
Ebury Publishing, a Random House Group company
First published in the USA by Harmony Books, an imprint of
Crown Publishing, a division of Random House Inc., in 2012

The Random House Group Limited Reg. No. 954009

Addresses for companies within the Random House Group can be found at:
www.randomhouse.co.uk

A CIP catalogue record for this book is
available from the British Library

The Random House Group Limited supports The Forest
Stewardship Council (FSC®), the leading international forest certification
organisation. Our books carrying the FSC label are printed on FSC® certified paper.
FSC is the only forest certification scheme endorsed by the leading environmental
organisations, including Greenpeace. Our paper procurement policy can be found at:
www.randomhouse.co.uk/environment

Printed and bound by CPI Group (UK) Ltd, Croydon, CR0 4YY

ISBN 9781846043666

Copies are available at special rates for bulk orders. Contact the sales development
team on 020 7840 8487 for more information.

To buy books by your favourite authors and register for offers, visit:
www.randomhouse.co.uk

To our wives and loving families

CONTENTS

Aristotle taught that the brain exists merely to cool the blood and is not involved in the process of thinking. This is true only of certain persons.

—Will Cuppy

PART 1

DEVELOPING YOUR GREATEST GIFT

A GOLDEN AGE
FOR THE BRAIN

What do we really know about the human brain? In the 1970s and 1980s, when the authors gained their training, the honest answer was "very little." There was a saying circulating back then: Studying the brain was like putting a stethoscope on the outside of the Astrodome to learn the rules of football.

Your brain contains roughly 100 billion nerve cells forming anywhere from a trillion to perhaps even a quadrillion connections called synapses. These connections are in a constant, dynamic state of remodeling in response to the world around you. As a marvel of nature, this one is minuscule and yet stupendous.

Everyone stands in awe of the brain, which was once dubbed "the three-pound universe." And rightly so. Your brain not only interprets the world, it creates it. Everything you see, hear, touch, taste, and smell would have none of those qualities without the brain. Whatever you experience today—your morning coffee, the love you feel for your family, a brilliant idea at work—has been specifically customized solely for you.

Immediately we confront a crucial issue. If your world is unique and customized for you and you alone, who is behind such remarkable creativity, you or the brain itself? If the answer is *you*, then the door to greater creativity is flung open. If the answer is *your brain*, then there may be drastic physical limitations on what you are able to achieve. Maybe your genes are holding you back, or toxic memories, or low self-esteem. Maybe you fall short because of limited expectations that have contracted your awareness, even though you don't see it happening.

The facts of the case could easily tell both stories, of unlimited potential or physical limitation. Compared with the past, today science is amassing new facts with astonishing speed. We have entered a golden age of brain research. New breakthroughs emerge every month, but in the midst of such exciting advances, what about the individual, the person who depends upon the brain for everything? Is this a golden age for *your* brain?

We detect an enormous gap between brilliant research and everyday reality. Another medical school saying from the past comes to mind: Each person typically uses only 10 percent of their brain. Speaking literally, that's not true. In a healthy adult, the brain's neural networks operate at full capacity all the time. Even the most sophisticated brain scans available would show no detectable difference between Shakespeare writing a soliloquy from *Hamlet* and an aspiring poet writing his first sonnet. But the physical brain is not nearly the whole story.

To create a golden age for your brain, you need to use the gift nature has given you in a new way. It's not the number of neurons or some magic inside your gray matter that makes life more vital, inspiring, and successful. Genes play their part, but your genes, like the rest of the brain, are also dynamic. Every day you step into the invisible firestorm of electrical and chemical activity that is the brain's environment. You act as leader, inventor, teacher, and user of your brain, all at once.

As leader, you hand out the day's orders to your brain.
As inventor, you create new pathways and connections
    inside your brain that didn't exist yesterday.

As teacher, you train your brain to learn new skills.

As user, you are responsible for keeping your brain in good
working order.

In these four roles lies the whole difference between the everyday brain—let's dub it the baseline brain—and what we are calling
super brain. The difference is immense. Even though you have not
related to the brain by thinking *What orders should I give today?* or
What new pathways do I want to create? that's precisely what you are
doing. The customized world that you live in needs a creator. The
creator isn't your brain; it's you.

Super brain stands for a fully aware creator using the brain to
maximum advantage. Your brain is endlessly adaptable, and you
could be performing your fourfold role—leader, inventor, teacher,
and user—with far more fulfilling results than you now achieve.

Leader: The orders you give are not just command prompts on a
computer like "delete" or "scroll to end of page." Those are mechanical commands built into a machine. Your orders are received by a
living organism that changes every time you send an instruction. If
you think *I want the same bacon and eggs I had yesterday,* your brain
doesn't change at all. If instead you think *What will I eat for breakfast
today? I want something new,* suddenly you are tapping into a reser-
voir of creativity. Creativity is a living, breathing, ever new inspiration that no computer can match. Why not take full advantage of
it? For the brain has the miraculous ability to give more, the more
you ask of it.

Let's translate this idea into how you relate to your brain now
and how you could be relating. Look at the lists below. Which do
you identify with?

BASELINE BRAIN

I don't ask myself to behave very differently today than I
did yesterday.

I am a creature of habit.

I don't stimulate my mind with new things very often.

I like familiarity. It's the most comfortable way to live.

If I'm being honest, there's boring repetition at home, work, and in my relationships.

SUPER BRAIN

I look upon every day as a new world.

I pay attention not to fall into bad habits, and if one sets in, I can break it fairly easily.

I like to improvise.

I abhor boredom, which to me means repetition.

I gravitate to new things in many areas of my life.

Inventor: Your brain is constantly evolving. This happens individually, which is unique to the brain (and one of its deepest mysteries). The heart and liver that you were born with will be essentially the same organs when you die. Not the brain. It is capable of evolving and improving throughout your lifetime. Invent new things for it to do, and you become the source of new skills. A striking theory goes under the slogan "ten thousand hours," the notion being that you can acquire any expert skill if you apply yourself for that length of time, even skills like painting and music that were once assigned only to the talented. If you've ever seen Cirque du Soleil, you might have assumed that those astonishing acrobats came from circus families or foreign troupes. In fact, every act in Cirque du Soleil, with few exceptions, is taught to ordinary people who come to a special school in Montreal. At one level, your life is a series of skills, beginning with walking, talking, and reading. The mistake we make is to limit these skills. Yet the same sense of balance that allowed you to toddle, walk, run, and ride a bicycle, given ten thousand hours (or less), can allow you to cross a tightrope strung between two skyscrapers. You are asking very little of your brain when you stop asking it to perfect new skills every day.

Which one do you identify with?

BASELINE BRAIN

I can't really say that I am growing as much as when I was younger.

If I learn a new skill, I take it only so far.

I am resistant to change and sometimes feel threatened by it.

I don't reach beyond what I am already good at.

I spend a good deal of time on passive things like watching television.

SUPER BRAIN

I will keep evolving my whole lifetime.

If I learn a new skill, I take it as far as I can.

I adapt quickly to change.

If I'm not good at something when I first try it, that's okay. I like the challenge.

I thrive on activity, with only a modicum of down time.

Teacher: Knowledge is not rooted in facts; it is rooted in curiosity. One inspired teacher can alter a student for life by instilling curiosity. You are in the same position toward your brain, but with one big difference: you are both student and teacher. Instilling curiosity is your responsibility, and when it comes, you are also the one who will feel inspired. No brain was ever inspired, but when you are, you trigger a cascade of reactions that light up the brain, while the incurious brain is basically asleep. (It may also be crumbling; there is evidence that we may prevent symptoms of senility and brain aging by remaining socially engaged and intellectually curious during our entire lifetime.) Like a good teacher, you must monitor errors, encourage strengths, notice when the pupil is ready for new challenges, and so on. Like a bright pupil, you must remain open to the things you don't know, being receptive rather than close-minded.

Which one do you identify with?

BASELINE BRAIN

I'm pretty settled in how I approach my life.
I am wedded to my beliefs and opinions.
I leave it to others to be the experts.
I rarely watch educational television or attend
   public lectures.
It's been a while since I felt really inspired.

SUPER BRAIN

I like reinventing myself.
I've recently changed a long-held belief or opinion.
There's at least one thing I am an expert on.
I gravitate toward educational outlets on television
   or in local colleges.
I'm inspired by my life on a day-to-day basis.

User: There's no owner's manual for the brain, but it needs nourishment, repair, and proper management all the same. Certain nutrients are physical; today a fad for brain foods sends people running for certain vitamins and enzymes. But the proper nourishment for the brain is mental as well as physical. Alcohol and tobacco are toxic, and to expose your brain to them is to misuse it. Anger and fear, stress and depression also are a kind of misuse. As we write, a new study has shown that routine daily stress shuts down the prefrontal cortex, the part of the brain responsible for decision making, correcting errors, and assessing situations. That's why people go crazy in traffic snarls. It's a routine stress, yet the rage, frustration, and helplessness that some drivers feel indicates that the prefrontal cortex has stopped overriding the primal impulses it is responsible for controlling. Time and again we find ourselves coming back to the same theme: Use your brain, don't let your brain use you. Road rage is an example of your brain using you, but so are toxic memories, the wounds of old traumas, bad habits you can't break, and most tragically, out-of-control addictions. This is a vastly important area to be aware of.

Which one do you identify with?

BASELINE BRAIN

I have felt out of control recently in at least one
  area of my life.
My stress level is too high, but I put up with it.
I worry about depression or am depressed.
My life can go in a direction I don't want it to.
My thoughts can be obsessive, scary, or anxious.

SUPER BRAIN

I feel comfortably in control.
I actively avoid stressful situations by walking away
  and letting go.
My mood is consistently good.
Despite unexpected events, my life is headed in
  the direction I want it to go.
I like the way my mind thinks.

Even though your brain doesn't come with an owner's manual, you can use it to follow a path of growth, achievement, personal satisfaction, and new skills. Without realizing it, you are capable of making a quantum leap in how you use your brain. Our final destination is the enlightened brain, which goes beyond the four roles you play. It is a rare kind of relationship, in which you serve as the observer, the silent witness to everything the brain does. Here lies transcendence. When you are able to be the silent witness, the brain's activity doesn't enmesh you. Abiding in complete peace and silent awareness, you find the truth about the eternal questions concerning God, the soul, and life after death. The reason we believe that this aspect of life is real is that when the mind wants to transcend, the brain is ready to follow.

A New Relationship

When Albert Einstein died in 1955 at the age of seventy-six, there was tremendous curiosity about the most famous brain of the twentieth century. Assuming that something physical must have created such genius, an autopsy was performed on Einstein's brain. Defying expectations that big thoughts required a big brain, Einstein's brain actually weighed 10 percent less than the average brain. That era was just on the verge of exploring genes, and advanced theories about how new synaptic connections are formed lay decades in the future. Both represent dramatic advances in knowledge. You can't see genes at work, but you can observe neurons growing new axons and dendrites, the threadlike extensions that allow one brain cell to connect with another. It's now known that the brain can form new axons and dendrites up to the last years of life, which gives us tremendous hope for preventing senility, for example, and preserving our mental capacity indefinitely. (So astounding is the brain's ability to make new connections that a fetus on the verge of being born is forming 250,000 new brain cells per minute, leading to millions of new synaptic connections per minute.)

Yet in so saying, we are as naïve as newspaper reporters waiting eagerly to tell the world that Einstein possessed a freakish brain—we still emphasize the physical. Not enough weight is given to how a person relates to the brain. We feel that without a new relationship, the brain cannot be asked to do new, unexpected things. Consider discouraged children in school. Such students existed in every classroom that all of us attended, usually sitting in the back row. Their behavior follows a sad pattern.

First the child attempts to keep up with other children. When these efforts fail, for whatever reason, discouragement sets in. The child stops trying as hard as the children who meet with success and encouragement. The next phase is acting out, making disruptive noises or pranks to attract attention. Every child needs attention, even if it is negative. The disruptions can be aggressive, but eventually the child realizes that nothing good is happening. Acting out leads to disapproval and punishment. So he enters the final phase, which is sullen silence. He makes no more effort to keep up in class. Other children mark him as slow or stupid, an outsider. School has turned into a stifling prison rather than an enriching place.

It's not hard to see how this cycle of behavior affects the brain. We now know that babies are born with 90 percent of their brains formed and millions of connections that are surplus. So the first years of life are spent winnowing out the unused connections and growing the ones that will lead to new skills. A discouraged child, we can surmise, aborts this process. Useful skills are not developed, and the parts of the brain that fall into disuse atrophy. Discouragement is holistic, encompassing brain, psyche, emotions, behavior, and opportunities later in life.

For any brain to operate well, it needs stimulation. But clearly stimulation is secondary to how the child feels, which is mental and psychological. A discouraged child relates to his brain differently than an encouraged child, and their brains must respond differently, too.

Super brain rests on the credo of connecting the mind and brain in a new way. It's not the physical side that makes the crucial difference. It's a person's resolve, intention, patience, hope, and diligence. These are all a matter of how the mind relates to the brain, for better or worse. We can summarize the relationship in ten principles.

A SUPER BRAIN CREDO
HOW THE MIND RELATES TO THE BRAIN

1. The process always involves feedback loops.
2. These feedback loops are intelligent and adaptable.
3. The dynamics of the brain go in and out of balance but always favor overall balance, known as homeostasis.
4. We use our brains to evolve and develop, guided by our intentions.
5. Self-reflection pushes us forward into unknown territory.
6. Many diverse areas of the brain are coordinated simultaneously.
7. We have the capacity to monitor many levels of awareness, even though our focus is generally confined to one level (i.e., waking, sleeping, or dreaming).
8. All qualities of the known world, such as sight, sound, texture, and taste, are created mysteriously by the interaction of mind and brain.
9. Mind, not the brain, is the origin of consciousness.
10. Only consciousness can understand consciousness. No mechanical explanation, working from facts about the brain, suffices.

These are big ideas. We have a lot of explaining to do, but we wanted you to see the big ideas up front. If you lifted just two words from the first sentence—*feedback loops*—you could mesmerize a medical school class for a year. The body is an immense feedback loop made up of trillions of tiny loops. Every cell talks to every other and listens to the answer it receives. That's the simple essence of feedback, a term taken from electronics. The thermostat in your living room senses the temperature and turns the furnace on if the room gets too cold. As the temperature rises, the thermostat takes in that information and responds by turning the furnace off.

The same back-and-forth operates through switches in the body that also regulate temperature. That's nothing fascinating, so far. But when you think a thought, your brain sends information to the heart, and if the message is one of excitement, fear, sexual arousal, or many other states, it can make the heart beat faster. The brain will send a countermessage telling the heart to slow down again, but if this feedback loop breaks down, the heart can keep racing like a car with no brakes. Patients who take steroids are replacing the natural steroids made by the endocrine system. The longer you take artificial steroids, the more the natural ones ebb, and as a result the adrenal glands shrink.

The adrenals are responsible for sending the message that slows down a racing heart. So if a patient stops taking a steroid drug all at once rather than tapering off, the body may be left with no brakes. The adrenal gland hasn't had time to regrow. In that event, somebody could sneak up behind you, yell "Boo!" and send your heart racing out of control. The result? A heart attack. With such possibilities, suddenly feedback loops start to become fascinating. To make them mesmerizing, there are extraordinary ways to use the brain's feedback. Any ordinary person hooked up to a biofeedback machine can quickly learn to control bodily mechanisms that usually run on automatic. You can lower your blood pressure, for example, or change your heart rate. You can induce the alpha-wave state associated with meditation and artistic creativity.

Not that a biofeedback machine is necessary. Try the following exercise: Look at the palm of your hand. Feel it as you look. Now imagine that it is getting warmer. Keep looking and focus on it getting warmer; see the color becoming redder. If you maintain focus on this intention, your palm will in fact grow warm and red. Tibetan Buddhist monks use this simple biofeedback loop (an advanced meditation technique known as *tumo*) to warm their entire bodies.

This technique is so effective that monks who use it can sit in freezing ice caves meditating overnight while wearing nothing more than their thin silk saffron robes. Now the simple feedback loop has become totally engrossing, because what we can induce merely by intending it may have no limit. The same Buddhist monks reach states of compassion, for example, that depend on physical changes in the prefrontal cortex of the brain. Their brains didn't do this on their own; they were following orders from the mind. Thus we cross a frontier. When a feedback loop is maintaining normal heart rhythm, the mechanism is involuntary—it is using you. But if you change your heart rate intentionally (for example, by imagining a certain someone who excites you romantically), you are using it instead.

Let's take this concept to the place where life can be miserable or happy. Consider stroke victims. Medical science has made huge advances in patient survival after even massive strokes, some of which can be attributed to better medications and to the upsurge of trauma units, since strokes are ideally dealt with as soon as possible. Quick treatment is saving countless lives, compared to the past.

But survival isn't the same as recovery. No drugs show comparable success in allowing victims to recover from paralysis, the most common effect of a stroke. As with the discouraged children, with stroke patients everything seems to depend on feedback. In the past they mostly sat in a chair with medical attention, and their course of least resistance was to use the side of the body that was unaffected by their stroke. Now rehabilitation actively takes the course of most resistance. If a patient's left hand is paralyzed, for example, the therapist will have her use only that hand to pick up a coffee cup or comb her hair.

At first these tasks are physically impossible. Even barely raising a paralyzed hand causes pain and frustration. But if the patient repeats the intention to use the bad hand, over and over, new feedback loops develop. The brain adapts, and slowly there is new function. We now see remarkable recoveries in patients who walk, talk, and use their limbs normally with intensive rehab. Even twenty years ago these functions would have languished or shown only minor improvements.

And all we have done so far is to explore the implications of two words.

The super brain credo bridges two worlds, biology and experience. Biology is great at explaining physical processes, but it is totally inadequate at telling us about the meaning and purpose of our subjective experience. What does it feel like to be a discouraged child or a paralyzed stroke victim? The story begins with that question, and biology follows second. We need both worlds to understand ourselves. Otherwise, we fall into the biological fallacy, which holds that humans are controlled by their brains. Leaving aside countless arguments between various theories of mind and brain, the goal is clear: We want to use our brains, not have them use us.

We'll expand on these ten principles as the book unfolds. Major breakthroughs in neuroscience are all pointing in the same direction. The human brain can do far more than anyone ever thought. Contrary to outworn beliefs, its limitations are imposed by us, not by its physical shortcomings. For example, when we were getting our medical and scientific training, the nature of memory was a complete mystery. Another saying circulated back then: "We know as much about memory as if the brain were filled with sawdust." Fortunately, brain scans were on the horizon, and today researchers can watch in real time as areas of the brain "light up," to display the firing of neurons, as subjects remember certain things. The Astrodome's roof is now made of glass, you could say.

But memory remains elusive. It leaves no physical traces in brain cells, and no one really knows how our memories are stored. But that's no reason to place any limitations on what our brains can remember. A young Indian math prodigy gave a demonstration in which she was asked to multiply two numbers, each thirty-two digits long, in her head. She produced the answer, which was sixty-four or -five digits long, within seconds of her hearing the two numbers. On average, most people can remember only six or seven digits at a glance. So what should be our norm for memory, the average person or the exceptional one? Instead of saying that the math prodigy has better genes or a special gift, ask another question: Did you train your brain to have a super memory? There are training courses for that skill, and average people who take them can perform feats like reciting the King James Bible from memory, using no more than the genes and gifts they were born with. Everything hinges on how you relate to your brain. By setting higher expectations, you enter a phase of higher functioning.

One of the unique things about the human brain is that it can do only what it thinks it can do. The minute you say, "My memory isn't what it used to be" or "I can't remember a thing today," you are actually training your brain to live up to your diminished expectations. Low expectations mean low results. The first rule of super brain is that your brain is always eavesdropping on your thoughts. As it listens, it learns. If you teach it about limitation, your brain will become limited. But what if you do the opposite? What if you teach your brain to be unlimited?

Think of your brain as being like a Steinway grand piano. All the keys are in place, ready to work at the touch of a finger. Whether a beginner sits down at the keyboard or a world-renowned virtuoso like Vladimir Horowitz or Arthur Rubinstein, the instrument is physically the same. But the music that comes out will be vastly different. The beginner uses less than 1 percent of the piano's potential; the virtuoso is pushing the limits of the instrument.

If the music world had no virtuosos, no one would ever guess at the amazing things a Steinway grand can do. Fortunately, research on brain performance is providing us with stunning examples of untapped potential brilliantly coming to life. Only now are these amazing individuals being studied with brain scans, which makes their abilities more astonishing and at the same time more mysterious.

Let's consider Magnus Carlsen, the Norwegian chess prodigy. He earned the highest ranking in chess, grand master, at the age of thirteen, the third youngest in history. Around that time, in a speed game, he forced Gary Kasparov, the former world chess champion, to a draw. "I was nervous and intimidated," Carlsen recalls, "or I would have beat him." To play chess at this level, a grand master must be able to refer, instantly and automatically, to thousands of games stored in his memory. We know the brain is not filled with sawdust, but how a person is able to recall such a vast storehouse of individual moves—amounting to many million possibilities—is totally mysterious. In a televised demonstration of his abilities, young Carlsen, who is now twenty-one, played ten opponents simultaneously in speed chess—with his back turned to the boards.

In other words, he had to keep in mind ten separate chess boards, with their thirty-two pieces, while the clock permitted only seconds for each move. Carlsen's performance defines the limit of memory, or a small slice of it. If it is difficult for a normal person to imagine having such a memory, the fact is that Carlsen isn't straining his brain. What he does, he says, feels completely natural.

We believe that every remarkable mental feat is a signpost showing the way. You won't know what your brain can do until you test its limits and push beyond them. No matter how inefficiently you are using your brain, one thing is certain: it is the gateway to your future. Your success in life depends on your brain, for the simple reason that all experience comes to us through our brains.

We want *Super Brain* to be as practical as possible, because it can solve problems that are far more difficult, or even impossible, for the baseline brain. Each chapter will end with its own Super Brain Solutions section, with a host of innovative suggestions for overcoming many of life's most common challenges.

FIVE MYTHS TO DISPEL

Relating to your brain in a new way is the way you can change reality. The more neuroscientists learn, the more it seems that the brain has hidden powers. The brain processes the raw material of life, as a servant to any desire you have, any vision you can imagine. The solid physical world cannot resist this power, and yet unlocking it requires new beliefs. Your brain cannot do what it thinks it cannot do.

Five myths in particular have proved limiting and obstructive to change. All were once accepted as fact, even a decade or two ago.

The injured brain cannot heal itself.
Now we know that the brain has amazing powers of healing, unsuspected in the past.

The brain's hardwiring cannot be changed.
In fact, the line between hard and soft wiring is shifting all the time, and our ability to rewire our brains remains intact from birth to the end of life.

Aging in the brain is inevitable and irreversible.

To counter this outmoded belief, new techniques for keeping the brain youthful and retaining mental acuity are arising every day.

The brain loses millions of cells a day, and lost brain cells cannot be replaced.

In fact, the brain contains stem cells that are capable of maturing into new brain cells throughout life. How we lose or gain brain cells is a complex issue. Most of the findings are good news for everyone who is afraid of losing mental capacity as they age.

Primitive reactions (fear, anger, jealousy, aggression) overrule the higher brain.

Because our brains are imprinted with genetic memory over thousands of generations, the lower brain is still with us, generating primitive and often negative drives like fear and anger. But the brain is constantly evolving, and we have gained the ability to master the lower brain through choice and free will. The new field of positive psychology is teaching us how best to use free will to promote happiness and overcome negativity.

It's good news that these five myths have been exploded. The old view made the brain seem fixed, mechanical, and steadily deteriorating. This turns out to be far from the case. You are creating reality at this very minute, and if that process remains alive and dynamic, your brain will be able to keep up with it, year after year.

Now let us discuss in detail how to dispel these old myths as they apply to your own experience and expectations.

Myth 1. The injured brain cannot heal itself.

When the brain is injured due to trauma in a car accident, for example, or due to a stroke, nerve cells and their connections to each other (synapses) are lost. For a long time it was believed that once the brain was injured, victims were stuck using whatever brain function they had left. But over the past two decades, a major discovery was made, and studies too numerous to count have confirmed it. When neurons and synapses are lost owing to injury, the neighboring neurons compensate for the loss and try to reestablish missing connections, which effectively rebuilds the damaged neural network.

The neighboring neurons step up their game and undergo "compensatory regeneration" of their main projecting parts (the main trunk, or axon, and the numerous threadlike branches, or dendrites). This new growth recoups the lost connections in the complex neural grid of which every brain cell is a part.

Looking back, we found it odd that science had once denied to brain cells an ability that was common to other nerves. Since the late 1700s, scientists had known that neurons in the peripheral nervous system (the nerves running through the body outside the brain and spinal cord) could regenerate. In 1776 William Cumberland Cruikshank, a Scottish-born anatomist, cut a half-inch section from the vagus or "wandering" nerve from a dog's neck. The vagus nerve runs to the brain along the carotid artery in the throat, and it is involved in regulating some major functions—heart rate, sweating, muscle movements for speech—and keeping the larynx open for breathing. If both branches of the nerve are cut, the result is lethal. Cruikshank cut only one branch and found that the gap he created was soon filled in with new nerve tissue. When he submitted his paper to the Royal Society, however, it met with skepticism and wasn't published for decades.

By then, other evidence was confirming that peripheral nerves like the vagus can heal when cut. (You can experience the same phenomenon if a deep gash leaves your finger numb; after a time feeling returns.) But for centuries people had believed that nerves in the central nervous system (the brain and spinal cord) lacked the same ability.

It's true that the central nervous system cannot regenerate with the same robustness and rapidity of the peripheral nervous system.

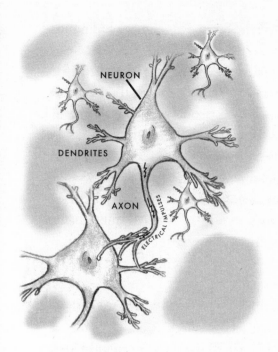

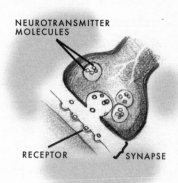

DIAGRAM 1: NEURONS AND SYNAPSES

Nerve cells (neurons) are true wonders of nature in their ability to create our sense of reality. Neurons connect to each other to form vast and intricate neural networks. Your brain contains over 100 billion neurons and up to a quadrillion connections, called *synapses.*

Neurons project wormlike threads known as axons and dendrites, which deliver both chemical and electrical signals across the gap between synapses. A neuron contains many dendrites to receive information from other nerve cells. But it has only one axon, which can extend out to over a meter (roughly 39 inches) in length. An adult human brain contains well over 100,000 miles of axons and countless dendrites—enough to wrap around the Earth over four times.

However, due to "neuroplasticity," the brain can remodel and remap its connections following injury. This remapping is the functional definition of neuroplasticity, which is now a hot-button issue. *Neuro* comes from *neuron*, while *plasticity* refers to being malleable. The old theory was that infants mapped their neural networks as a natural part of their development, after which the process stopped and the brain became hardwired. We now view the projections of nerve cells in the brain like long thin worms continually reconfiguring themselves in response to experience, learning, and injury. To heal and to evolve are intimately linked.

Your brain is remodeling itself right now. It doesn't take an injury to trigger the process—being alive is enough. You can promote neuroplasticity, moreover, by exposing yourself to new experiences. Even better is to deliberately set out to learn new skills. If you show passion and enthusiasm, all the better. The simple step of giving an older person a pet to take care of instills more willingness to live. The fact that the brain is being affected makes a difference, but we need to remember that neurons are servants. The dissecting knife reveals changes at the level of nerve projections and genes. What really invigorates an older person, though, is acquiring a new purpose and something new to love.

Neuroplasticity is better than mind over matter. It's mind turning into matter as your thoughts create new neural growth. In the early days, the phenomenon was scoffed at and neuroscientists were belittled for using the term *neuroplasticity*. Still, many new concepts that will likely be seminal and mainstream decades from now are today judged meaningless and useless. Neuroplasticity overcame a rough start to become a star.

That mind has such power over matter was momentous for both of us in the 1980s. Deepak was focused on the spiritual side of the mind-body connection, promoting meditation and alternative medicine. He was inspired by a saying he ran across early on: "If you want to know what your thoughts were like in the past, look at your body today. If you want to know what your body will be like in the future, look at your thoughts today."

For Rudy, this paradigm-breaking discovery really hit home when he was a graduate student at Harvard Medical School in the neuroscience program. Working at Boston Children's Hospital, he was trying to isolate the gene that produces the main brain toxin in Alzheimer's disease, the amyloid beta protein—the A beta peptide for short—the sticky substance that accumulates in the brain and correlates with neurons becoming dysfunctional and breaking down. Rudy was furiously poring over every paper he could find on Alzheimer's and this toxic amyloid. It can take the form of the beta-amyloid in Alzheimer's disease, or the prion amyloid in Mad Cow–related diseases.

One day he read a paper showing how the brain of an Alzheimer's patient had dealt with the accumulation of beta-amyloid in an effort to remodel the stricken part of the brain responsible for short-term memory, the hippocampus, which is located in the temporal lobe (so called because it is located in the skull beneath the temples).

The fact that the brain could try to find a way to bypass devastating damage changed Rudy's entire view of the disease he had been studying day and night in a snug lab the size of a small supply room on the fourth floor of the hospital. Between 1985 and 1988, he focused on identifying the gene that makes beta-amyloid accumulate excessively in the brains of Alzheimer's patients. Every day he worked side by side with his colleague Rachel Neve, while in the background a music soundtrack played, especially by Keith Jarrett, arguably the best jazz pianist who has ever lived.

Rudy loved Keith Jarrett's concerts for their brilliant improvisation. Jarrett had his own word for it: "extemporized." In other words, they were on the spot, radically spontaneous. To Rudy, Jarrett expressed in music the way the brain works in the everyday world—responding in the moment in creative directions based on the foundation of a lifetime's worth of experiences. Wisdom renewing itself in the moment. Memory finding fresh life. It is fair to say that when Rudy discovered the first Alzheimer's gene, the amyloid precursor protein (APP) in that small fourth-floor lab, his muse was Keith Jarrett.

Against this background enters the paper in 1986 that gave hope for Alzheimer's patients to regenerate brain tissue. It was an unseasonably cold day even for a Boston winter, and Rudy was sitting in the open stacks on the third floor of the library at Harvard Medical School, breathing the familiar scent of old musty paper—some of these scientific papers hadn't seen the light of day for decades.

Among the new articles on Alzheimer's was one in the journal *Science*, reported by Jim Geddes and colleagues, with the intriguing title "Plasticity of Hippocampal Circuitry in Alzheimer's Disease." After glancing through it, Rudy sprinted to the change machine to get a handful of dimes for the copy machine. (The luxury of computerized journals was still in the future.) After carefully reading it together with Rachel, they stared at each other wide-eyed for what seemed hours, finally exclaiming, "How cool is that!?" The mystery of a brain that could heal itself had entered their lives.

The essence of that seminal study was this. In Alzheimer's disease, one of the first things that goes wrong is short-term memory. In the brain, the key neural projections that allow sensory information to be stored are literally severed. (We are in the same field as Cruikshank when he cut a dog's vagus nerve.) More specifically, there is a small swollen bag of nerve cells in the brain called the entorhinal cortex, which acts as a way station for all the sensory information you take in, relaying it on to the hippocampus for short-term storage. (If you can remember that Rudy was working with a colleague named Rachel, that's the hippocampus doing its job.) The hippocampus takes its name from the Latin word for seahorse, which it resembles. Make two C's out of your thumb and forefinger on each hand facing each other and then interlock them in a parallel plane, and that is roughly the right shape.

Let's say you come home from shopping and want to tell a friend about some red shoes that would be perfect for her. The image of those shoes, passing through the entorhinal cortex, is relayed via neural projections called the perforant pathway. Now we have arrived at the physiological reason why someone with Alzheimer's will not remember those shoes. In Alzheimer's patients the exact region where the perforant pathway pierces the hippocampus routinely contains an abundance of neurotoxic beta-amyloid, which short-circuits the transfer of sensory information. Adding to the damage, nerve endings begin to shrink and break down in the same region, effectively severing the perforant pathway.

The nerve cells in the entorhinal cortex that should be sprouting those nerve endings soon die because they rely on growth factors, the proteins that support their survival, to be shunted up the nerve endings that once connected to the hippocampus. Eventually, the person can no longer achieve short-term memory and learning, and dementia sets in. The result is devastating. As one saying goes, you don't know you have Alzheimer's because you forget where you put your car keys. You know you have Alzheimer's when you forget what they are for.

In his seminal study, Geddes and his colleagues showed that in this area of massive neuronal demise, something nothing short of the magical occurs. The surviving neighboring neurons begin to sprout new projections to compensate for the ones that were lost. This is a form of neuroplasticity called compensatory regeneration. For the first time, Rudy was encountering one of the most miraculous properties of the brain. It was as if a rose were plucked from a bush, and the bush next to it handed it a new rose.

Rudy suddenly had a deep appreciation for the exquisite power and resilience of the human brain. Never count the brain out, he thought. With neuroplasticity, the brain has evolved into a marvelously adaptable and remarkably regenerative organ. Hope existed that even in a brain being damaged by Alzheimer's, one need only catch it early enough, and neuroplasticity may be triggered. It's one of the brightest possibilities for future research.

Myth 2. The brain's hardwiring cannot be changed.

During all the time before neuroplasticity was proved to be legitimate, medicine could have listened to the Swiss philosopher Jean-Jacques Rousseau, who argued in the middle 1700s that nature was not stagnant or machinelike but alive and dynamic. He went on to propose that the brain was continually reorganized according to our experiences. Therefore, people should practice mental exercise the same as physical exercise. For all intents and purposes, this may have been the first declaration that our brains are flexible and plastic, capable of adapting to changes in our environment.

Much later, in the middle of the twentieth century, American psychologist Karl Lashley provided evidence for this phenomenon. Lashley trained rats to seek out food rewards in a maze and then removed large portions of their cerebral cortex, bit by bit, to test when they would forget what they had previously learned. He assumed, given how delicate brain tissue is and how totally dependent a creature is on its brain, that removing a small portion would lead to severe memory loss.

Shockingly, Lashley found that he could take out 90 percent of a rat's cortex, and the animal still successfully navigated the maze. As it turned out, in learning the maze, the rats create many different types of redundant synapses based on all their senses. Many different parts of their brains interact to form a variety of overlapping sensory associations. In other words, the rats were not just seeing their way to the food in the maze; they were smelling and feeling their way as well. When bits of the cerebral cortex were removed, the brain would sprout new projections (axons) and form new synapses to take advantage of other senses, using the cues that remained, however tiny.

Here we have the first strong clue that "hardwiring" should be greeted with skepticism. The brain has circuitry but no wires; the circuits are made of living tissue. More important, they are reshaped by thoughts, memories, desires, and experiences. Deepak remembers a controversial medical article from 1980 entitled, half in jest, "Is the Brain Really Necessary?" It was based on the work of British neurologist John Lorber, who had been working with victims of a brain disorder known as hydrocephalus ("water on the brain"), in which excessive fluid builds up. The pressure that results squeezes the life out of brain cells. Hydrocephalus leads to retardation as well as other severe damage and even death.

Lorber had previously written about two infants born with no cerebral cortex. Yet despite this rare and fatal defect, they seemed to be developing normally, with no external signs of damage. One child survived for three months, the other for a year. If this were not remarkable enough, a colleague at Sheffield University sent Lorber a young man who had an enlarged head. He had graduated from college with a first-class honors degree in mathematics and had an IQ of 126. He had no symptoms of hydrocephalus; the young man was leading a normal life. Yet a CAT scan revealed, in Lorber's words, that he had "virtually no brain." The skull was lined with a thin layer of brain cells about a millimeter thick (less than one-tenth of an inch), while the rest of the space in the skull was filled with cerebral fluid.

This is an appalling disorder to contemplate, but Lorber pushed on, recording more than six hundred cases. He divided his subjects into four categories depending on how much fluid was in the brain. The most severe category, which accounted for only 10 percent of the sample, consisted of people whose brain cavity was 95 percent filled with fluid. Of these, half were severely retarded; the other half, however, had IQs over 100.

Not surprisingly, skeptics went on the attack. Some doubters said that Lorber must not have read the CAT scans correctly, but he assured them that his evidence was solid. Others argued that he hadn't actually weighed the brain matter that remained, to which he drily replied, "I can't say whether the mathematics student has a brain weighing 50 grams or 150 grams, but it is clear that it is nowhere near the normal 1.5 kilograms." In other words, 2 to 6 ounces may be involved, but that's nowhere near 3 pounds. More sympathetic neurologists declared that these results were proof positive of how redundant the brain is—many functions are copied and overlap. But others shrugged off this explanation, noting that "redundancy is a cop-out to get around something you don't understand." To this day, the whole phenomenon is shrouded in mystery, but we need to keep it in mind as our discussion unfolds. Could this be a radical example of the mind's power to have the brain—even a drastically reduced brain—carry out commands?

But we must consider more than brain injury. In a more recent example of neural rewiring, neuroscientist Michael Merzenich and colleagues at the University of California, San Francisco, took seven small monkeys who were trained to use their fingers to find food. The setup was that small banana-flavored pellets were placed at the bottom of small compartments, or food wells, in a plastic board. Some of the wells were wide and shallow; others were narrow and deep. Naturally when a monkey tried to retrieve the food, it would be more successful with the wide, shallow wells and fail at the narrow, deep ones, more often than not. However, as time went on, all the monkeys became extremely skillful, and eventually they succeeded every time, no matter how far their little fingers had to reach to retrieve a pellet.

The team then took brain scans of a specific area known as the somatosensory cortex, which controls the movement of fingers, hoping to show that the experience of learning a skill had actually altered the monkeys' brains. It was a success. This brain region rewired itself to other regions in order to increase the odds of finding more food in the future. Merzenich argued that as brain regions begin to newly interact, rewiring creates a new circuit. In this form of neuroplasticity, "neurons that fire together, wire together." In our everyday lives, if we intentionally set out to learn new things or do familiar things in new ways (such as commuting to work via a new route or taking the bus instead of a car), we effectively rewire our brains and improve them. A physical workout builds muscle; a mental workout creates new synapses to strengthen the neural network.

Many other examples reinforce the idea that the traditional doctrine of the stagnant, unchanging brain was false. Stroke patients did not have to be stuck with the brain damage caused by a broken blood vessel or clot. As brain cells die, the neighboring cells can compensate, maintaining the integrity of the neural circuit. To make this more personal, you see the house you grew up in, remember your first kiss, and cherish your circle of friends thanks to a highly personalized neural circuit that took a lifetime to create.

One example of the miraculous ability of the brain to rewire itself is the case of an auto mechanic who suffered severe brain trauma after being thrown from his car in a traffic accident. He was paralyzed and able only to eye-blink or slightly nod his head to communicate. After seventeen years, however, this man spontaneously bounced out of his semicomatose condition. In the week following, he underwent an astonishing recovery, to the point of regaining fluent speech and some movement in his limbs. Over the next year and a half brain imaging gave visible evidence that he was regenerating new pathways that could restore his brain function. The healthy nerve cells were sprouting new axons (main trunks) and dendrites (numerous threadlike branches) to create neural circuitry that would compensate for the dead nerve cells—classic neuroplasticity!

The bottom line is that we are not "hardwired." Our brains are incredibly resilient; the marvelous process of neuroplasticity gives you the capability, in your thoughts, feelings, and actions, to develop in any direction you choose.

Myth 3. Aging in the brain is inevitable and irreversible.

A movement known as the new old age is sweeping society. The social norm for the elderly used to be passive and grim; consigned to rocking chairs, they were expected to enter physical and mental decline. Now the reverse is true. Older people have higher expectations that they will remain active and vital. As a result, the definition of old age has shifted. A survey asked a sample of baby boomers "When does old age begin?" The average answer was 85. As expectations rise, clearly the brain must keep pace and accommodate the new old age. The old theory of the fixed and stagnant brain held that an aging brain was inevitable. Supposedly brain cells died continuously over time as a person aged, and their loss was irreversible.

Now that we understand how flexible and dynamic the brain is, the inevitability of cell loss is no longer valid. In the aging process—which progresses at about 1 percent a year after the age of thirty—no two people age alike. Even identical twins, born with the same genes, will have very different patterns of gene activity at age seventy, and their bodies can be dramatically different as a result of lifestyle choices. Such choices didn't add or subtract from the genes they were born with; rather, almost every aspect of life—diet, activity, stress, relationships, work, and the physical environment—changed the activity of those genes. Indeed, no single aspect of aging is inevitable. For any function, mental or physical, you can find people who improve over time. There are ninety-year-old stockbrokers who conduct complex transactions with memories that have improved over time.

The problem is that too many of us adhere to the norm. As we get older, we tend to get lazy and apathetic about learning. It takes smaller stresses to upset us, and these stresses linger for a longer time. What used to be dismissed as an elderly person's "being set in his ways" can now be traced to the mind-brain connection. Sometimes the brain is dominant in this partnership. Suppose a restaurant is behind in seating its patrons who have reservations. A younger person who must stand in line feels mild annoyance, but it dissipates once he is seated. An older person may react with a flash of anger—and remain resentful even after he has been seated. This is the differ-

ence in the physical stress response that the brain is responsible for.
Likewise, when older people get overwhelmed by too much sensory
input (a noisy traffic jam, a crowded department store), their brains
are probably exhibiting diminished function to take in tidal waves of
data from the busy world.

Much of the time, however, the mind dominates the mind-brain
connection. As we get older, we tend to simplify our mental activities, often as a defense mechanism or security blanket. We feel
secure with what we know, and we go out of our way to avoid learning anything new. The behavior strikes younger people as irritabil-
ity and stubbornness, but the real cause can be traced to the dance
between mind and brain. For many but not all older people, the
music slows down. What's most important is that they not walk
off the dance floor—which would pave the way for decline of both
mind and brain. Instead of your brain making new synapses, it keeps
hardwiring the ones you already have. In this downward spiral of
mental activity, the aged person will eventually have fewer dendrites
and synapses per neuron in the cerebral cortex.

Fortunately, conscious choices can be made. You can choose to
be aware of the thoughts and feelings being evoked in your brain at
every minute. You can choose to follow an upward learning curve no
matter how old you are. By doing so, you will create new dendrites,
synapses, and neural pathways that enhance the health of your brain
and even help stave off Alzheimer's disease (as suggested by the latest research findings).

If inevitability has been called into question, what about the ir-
reversibility of the effects of aging? As we get older, many of us
increasingly feel that our memories are going downhill. We cannot remember why we entered a room and joke, rather defensively,
about having senior moments. Rudy has a wonderful cat that follows him everywhere like a dog. More than once, Rudy has gotten
up from his chair in the living room and headed for the kitchen
with the cat in tow, only to find, when he gets there, that he and

the cat are staring blankly at each other. Neither of them knows the purpose of the journey. While we may refer to these lapses as instances of age-related memory loss, they are actually due to a lack of learning—registering new information in the brain. In many cases, we become so jaded or distracted about what we are doing that simple attention deficit leads to lack of learning. When we cannot remember a simple fact like where we put our keys, it means we did not learn or register where we put them in the first place. As users of our brains, we didn't record or consolidate the sensory information into a short-term memory during the process of putting down the keys. One cannot *remember* what one never *learned*.

If you remain alert, a healthy brain will continue to serve you as you age. You should expect alertness, rather than dread of impairment and senility. In our view—Rudy speaks as a leading researcher on Alzheimer's—a public campaign that created alarm about senility would have a damaging effect. Expectations are powerful triggers for the brain. If you expect to lose your memory and notice every minor lapse with anxiety, you are interfering with the natural, spontaneous, and effortless act of remembering. Biologically, up to 80 percent of people over seventy do not have significant memory loss. Our expectations should follow that finding, rather than our hidden and largely unfounded dread.

If you become apathetic and jaded about your life, or if you simply become less enthusiastic about your moment-to-moment experiences, your learning potential is impaired. As physical evidence, a neurologist can point to the synapses that must be consolidated for short-term memory. But in most cases a mental event has preceded the physical evidence: we never learned what we believe we have forgotten.

Nothing solidifies a memory like emotion. When we are children, we learn effortlessly because the young are naturally passionate and enthusiastic about learning. Emotions of joy and wonder, but also of horror and dread, intensify learning. That locks memories in, often for life. (Try to remember your first hobby or your first kiss. Now try to remember the first congressman you voted for, or the make of your neighbor's car when you were ten. Usually the one is easy and the other not so easy—unless you had an early passion for politics and cars.)

Sometimes the wow factor that works for children also works for adults. Strong emotion is often the key. We all remember where we were when the 9/11 attacks occurred, just as older people remember where they were on April 12, 1945, when President Roosevelt died suddenly on vacation at "the little White House" in Warm Springs, Georgia. Since memory remains so uncharted, we can't say, in terms of brain function, why intense emotions can cause highly detailed memories to be deposited. Some intense emotions may have the opposite effect: in childhood sex abuse, for example, that powerful trauma is suppressed and can be retrieved only with intensive therapy or hypnosis. These matters can't be resolved until some basic questions are answered: What is a memory? How does the brain store a memory? What kind of physical trace, if any, does a memory leave inside a brain cell?

Until answers arise, we believe that behavior and expectations are key. When you become passionate and excited about learning again, the way children are, new dendrites and synapses will form, and your memory can once again be as strong as it was when you were younger. As well, when you recall an old memory through active retrieval (i.e., you search your mind to recall the past accurately), you make new synapses, which strengthens old synapses, increasing the odds that you will recall the same memory again in the future. The onus is on you, the brain's leader and user. You are not your brain; you are much more. In the end, that's the one thing always worth remembering.

Myth 4. The brain loses millions of cells a day, and lost brain cells cannot be replaced.

The human brain loses about 85,000 cortical neurons per day, or about one per second. But this is an infinitesimal fraction (0.0002 percent) of the roughly 40 billion neurons in your cerebral cortex. At this rate, it would take more than six hundred years to lose half of the neurons in your brain! We have all grown up being told that once we lose brain cells, they are gone forever and never replaced. (In our adolescence, this warning was a standard part of parental lectures about the dangers of alcohol.) Over the past several decades, however, permanent loss has been shown not to be the case. Researcher Paul Coleman, at the University of Rochester, showed that the total number of nerve cells in your brain at age twenty does not significantly change when you reach seventy.

The growth of new neurons is called neurogenesis. It was first observed about twenty years ago in the brains of certain birds. For example, when zebra finches are developing and learning new songs for purposes of mating, their brains grow remarkably in size—new nerve cells are produced to accelerate the learning process. After a finch learns the song, many of the new nerve cells die off, returning the brain to its original size. This process is known as programmed cell death, or apoptosis. Genes not only know when it is time for new cells to be born (say, when we grow permanent teeth to replace baby teeth or undergo the changes of puberty) but also when it is time for a cell to die as when we slough off skin cells, lose our blood corpuscles after a few months, and many other cases. Most people are surprised to learn this fact. Death exists in the service of life—you may resist the idea, but your cells understand it completely.

In the decades following these seminal discoveries, researchers observed neurogenesis in the mammalian brain, particularly in the hippocampus, which is used for short-term memory. We now know that several thousand new nerve cells are born in the hippocampus every day. Neuroscientist Fred Gage at the Salk Institute showed that physical exercise and environmental enrichment (stimulating surroundings) stimulate the growth of new neurons in mice. One sees the same principle at work in zoos. Gorillas and other primates languish if they are kept in confined cages with nothing to do, but they flourish in large enclosures with trees, swings, and toys. If we could learn exactly how to safely induce neurogenesis in the human brain, we could more effectively treat conditions where brain cells have been lost or severely damaged: Alzheimer's disease, traumatic brain injury, stroke, and epilepsy. We could also reliably maintain the health of our brains as we age.

Alzheimer's researcher Sam Sisodia at the University of Chicago showed that physical exercise and mental stimulation protect mice from getting Alzheimer's disease, even when they have been engineered to carry a human Alzheimer's mutation in their genome. Other studies in rodents offer encouragement for the normal brain, too. By choosing to exercise every day, you can increase the number of new nerve cells, just as you do when you actively seek to learn new things. At the same time, you promote the survival of these new cells and connections. In contrast, emotional stress and trauma leads to the production of glucocorticoids in the brain, toxins that inhibit neurogenesis in animal models.

We can safely discard the myth about losing millions of brain cells a day. Even the parental warning that alcohol kills off brain cells has turned out to be a half-truth. Casual alcohol use actually kills only a minimal number of brain cells, even among alcoholics (who, however, incur many real health dangers). The actual loss from drinking occurs in dendrites, but studies seem to indicate that this damage is mostly reversible. The bottom line for now is that as we age, key areas of the brain involved with memory and learning continue to produce new nerve cells, and that this process can be stimulated by physical exercise, mentally stimulating activities (like reading this book), and social connectedness.

Myth 5. Primitive reactions (fear, anger, jealousy, aggression) overrule the higher brain.

Most people have at least caught some wind that the first four myths are untrue. The fifth myth, however, seems to be gaining ground.

The rationale for declaring that human beings are driven by primitive impulses is partly scientific, partly moral, and partly psychological. To put it in a sentence, "We were born bad because God is punishing us, and even science agrees." Too many people believe some part of this sentence, if not all of it.

Let's examine what seems to be the rational position, the scientific argument. All of us are born with genetic memory that provides us with the basic instincts we need to survive. Evolution aims to ensure the propagation of our species. Our instinctive needs work hand in hand with our emotional urges to gather food, find shelter, seek power, and procreate. Our instinctive fear helps us avoid dangerous situations that threaten the lives of ourselves and our kin.

Thus an evolutionary argument is used to persuade us that our fears and desires, instinctively programmed in us from the womb, are in charge, overruling our higher, more evolved brain, with its reason and logic (glossing over the all-too-obvious irony that the higher brain invented the theory that demoted it). Undoubtedly, instinctive reactions are built into the brain's structure. Some neuroscientists find convincing the argument that certain people are programmed to become antisocial, criminals, or rage-aholics, much as others are programmed for anxiety, depression, autism, and schizophrenia.

But emphasizing the lower brain overlooks a powerful truth. The brain is multidimensional, in order to allow *any* experience to occur. Which experience will dominate is neither automatic nor genetically programmed. There is a balance between desire and restraint, choice and compulsion. Accepting that biology is destiny defeats the whole purpose of being human: we should submit to destiny only as a last, desperate choice, but the argument for a domineering lower brain makes submission the first choice. How can that be condoned? We shrug that our forebears resigned themselves to human wrongdoing because it was said to be inherited from Adam and Eve's disobedience in the Garden of Eden. Genetic inheritance runs the danger of inducing the same resignation, dressed up in scientific garb.

Even though we experience fear and desire every day as natural reactions to the world, we do not have to be ruled by them. A frustrated driver stalled on the L.A. freeway in choking smog will feel the same fight-or-flight response as his ancestors hunting antelope on the African savannah or saber-toothed tigers in northern Europe. This response to stress, an instinctual drive, was built into us, but it doesn't make drivers abandon their vehicles en masse to run away or attack each other. Freud held that civilization depends on our overriding primitive urges so that higher values can prevail, which sounds true enough. But he believed pessimistically that we pay a high price for it. We repress our lower drives but never extinguish them or make peace with our deeper fears and aggressions. The result is eruptions of mass violence like the two world wars, when all of that repressed energy takes its toll in horrendous, uncontrollable ways.

We can't summarize the thousands of books that have been written on this subject, or offer the perfect answer. But surely to label human beings as puppets of animal instinct is wrong, in the first place because it is so unbalanced. The higher brain is just as legitimate, powerful, and evolutionary as the lower brain. The largest circuits in the brain, which form feedback loops between the higher and lower areas, are malleable. If you are an enforcer in professional hockey and your job is to start fights on the ice, you've probably chosen to shape your brain circuitry to favor aggression. But it was always a choice, and if the day ever comes when you regret your choice, you can retire to a Buddhist monastery, meditate upon compassion, and shape the brain's circuitry in a new, higher direction. The choice is always there.

With rare exception, freedom of choice is not prohibited by preset programming. *My brain made me do it* has become a default explanation for almost every undesirable behavior. We can be consciously aware of our emotions and choose not to identify with them. This is more easily said than done for a person suffering from bipolar disorder, drug addiction, or a phobia. But the road to brain well-ness begins with awareness. It also ends in awareness, and awareness allows every step along the way. In the brain, energy flows where awareness goes.

When the energy stops flowing, you become stuck. Stuckness is an illusion, but when it is happening to you, it feels very real. Consider someone who is deathly afraid of spiders. Phobias are fixed (i.e., stuck) reactions. An arachnophobe cannot see a spider without an automatic rush of fear. The lower brain triggers a complex chemical cascade. Hormones race through the bloodstream to speed up the heart and raise blood pressure. Muscles prepare for fight or flight. The eyes become tightly focused, with tunnel vision on the thing one fears. The spider becomes enormous in the mind's eye. So powerful is the fear reaction that the higher brain—the part that knows how small and harmless most spiders are—gets blacked out.

Here is a prime example of the brain using you. It imposes a false reality. All phobias are distortions of reality at bottom. Heights are not automatically a cause for panic; nor are open spaces, flying in an airplane, and the myriad other things that phobics are afraid of. By giving up the power to use their brains, phobics become stuck in a fixed reaction.

Phobias can be successfully treated by bringing in awareness and restoring control to the user of the brain, where it belongs. One technique is to have the person imagine what he is afraid of. An arachnophobe, for example, is asked to see a spider and to make the image grow bigger and smaller. Then to cause the image to move back and forth. This simple act of giving motion to the feared object can be very effective in dispelling its power to induce, because fear freezes the mind. Gradually, the therapy can move to a spider in a glass box. The phobic is asked to move as close as he can without feeling panicky. The distance is allowed to change depending on his comfort level, and in time this freedom to change also restores control. The phobic learns that he has more choices than simply running away.

Obviously, the higher brain can override even the most instinc-tual fears; otherwise, we wouldn't have mountain climbers (fear of heights), tightrope walkers (fear of falling), and lion tamers (fear of death). The unhappy fact, however, is that we are all like the phobic who cannot even imagine the picture of a spider without breaking out in a cold sweat. We surrender to fears, not of spiders, but of what we call normal: failure, humiliation, rejection, old age, sickness, and death. It's tragically ironic that the same brain that can conquer fear should also subject us to fears that haunt us all our lives.

So-called lower creatures enjoy freedom from psychological fear. When a cheetah attacks a gazelle, it panics and fights for its life. But if no predator is present, the gazelle leads an untroubled life, so far as we know. We humans, however, suffer terribly in our inner world, and this suffering gets translated into physical problems. The stakes are very high when it comes to letting your brain use you. But if you start to use it instead, the rewards are unlimited.

SUPER BRAIN SOLUTIONS

MEMORY LOSS

We've been pushing the theme that you need to relate to your brain in a new way. This especially holds true for memory. We cannot expect memory to be perfect, and how you respond to its imperfections is up to you. If you see every little lapse as a warning sign of inevitable decline with age, or an indication that you lack intellect, you are stacking the odds to make your belief come true. Every time you complain "My memory is going," you reinforce that message in your brain. In the balance of mind and brain, most people are too quick to blame the brain. What they should be looking at is habit, behavior, attention, enthusiasm, and focus, all of which are primarily mental.

Once you stop paying attention and give up on learning new things, you give memory no encouragement. A simple axiom holds: whatever you pay attention to grows. So to encourage your memory to grow, you need to pay attention to how your life is unfolding. What does this mean, specifically? The list is long, but it contains activities that come naturally. The only difference as you age is that you have to make more conscious choices than you did earlier in life:

A MINDFUL MEMORY PROGRAM

Be passionate about your life and the experiences you fill it with.

Enthusiastically learn new things.

Pay attention to the things you will need to remember later. Most memory lapses are actually learning lapses.

Actively retrieve older memories; rely less on memory crutches like lists.

Expect to keep your memory intact. Resist lower expectations from people who rationalize memory loss as "normal."

Don't blame or fear occasional lapses.

If a memory doesn't come immediately, don't brush it off as lost. Be patient and take the extra seconds for the brain's retrieval system to work. Focus on things or people you associate with the lost memory, and you will likely recall it. All memories are associated with other earlier ones. This is the basis of learning.

Be wide-ranging in your mental activities. Doing a crossword puzzle uses a different part of the memory system than remembering what groceries you need, and both are different from learning a new language or recalling the faces of people just met. Actively exercise all aspects of memory, not just the ones that come most easily.

The common thread in this program is to keep up the mind-brain connection. Every day counts. Your brain never stops paying attention to what you tell it, and it can respond very quickly. A longtime friend of Deepak's, a medical editor, has prided himself on his memory since childhood. As he is quick to point out, he doesn't have a photographic (or eidetic) memory. Instead, he "keeps his antennae out," as he describes it. As long as he keeps paying attention to his day-to-day existence, he can retrieve memories quickly and reliably.

Recently this man turned sixty-five, as did most of his friends. They began to exchange wry jokes about their senior moments. (Sample: "My memory is as good as it ever was. I just don't have same-day delivery.") The man began to notice random lapses in himself, although he had no trouble using his memory when he did research for his work.

"Without really worrying about it," he says, "I decided to start making a grocery list. Up to then, I'd never made any lists. I went out shopping and simply remembered what I wanted. This was true even if I had to stock my depleted kitchen with several bags of groceries.

"I started keeping a grocery list on my desktop, and an amazing thing happened. Within a day or two I couldn't remember what I wanted to buy. Without my list in hand, I was helpless, wandering the aisles of the grocery store in the hopes that once I spied potatoes or maple syrup, I'd remember that it was what I came for.

"At first I laughed it off, until one week when I forgot to buy sugar on two visits to the supermarket. Now I'm trying to wean myself off the list. I still intend to, but you get dependent on lists very quickly."

Learning from his example, sit down and consider the things you could be paying more attention to while using fewer crutches. Our Mindful Memory Program will guide you, since it includes the major areas where it pays to pay attention. The most familiar things may seem unimportant, but they count.

Can you wean yourself off making lists for things that you can remember? Try taking your grocery list to the supermarket but not looking at it. Buy as much as you can from memory, and only then consult your list. When you get to the point that you leave nothing out, wean yourself from the list entirely.

Can you stop blaming yourself for memory lapses? Catch yourself the next time you would automatically say "I can't remember a thing" or "Another senior moment." Be patient and wait. If you expect memories to come, they almost always do.

Stop blocking your memory. Retrieving a memory is delicate: you can easily step in the way of remembering by being busy, distracted, worried, stressed out, tired from lack of sleep, or overtaxed mentally from doing two or more things at once. Examine these things first, before you blame your brain.

Set up an environment that's good for memory, one that has the opposite of what we just mentioned as obstacles. In other words, take care of stress, get enough sleep, be regular in your habits, don't overtax yourself mentally with multitasking, and so forth. Developing regular habits helps, since the brain operates more easily on repetition. If you live in a scattered and distracted way, the sensory overload to your brain is damaging and unnecessary.

If you are getting older and feel that memory loss could be occurring, don't panic or resign yourself to the inevitable. Instead, focus your effort on mental activity that boosts brain function. Certain software, including so-called "brain gyms," and books like *Neurobics,* coauthored by the Duke University neurobiologist Larry Katz, are designed to exercise the brain in a systematic way. The reports of reversing mild-to-moderate memory loss by exercising the brain are as yet anecdotal, but they are encouraging nevertheless.

Finally, look upon this whole project as natural. Your brain was designed to follow your lead, and the more relaxed you are, the better that will be for your mind-brain partnership. The best memory is one you rely upon with simple confidence.

HEROES
OF SUPER BRAIN

Now that we have dispelled some false myths, the path to super brain looks clearer. But a new obstacle up ahead is blocking the way: complexity. The neural network of your brain is the computer of your body, but it is also the computer of your life. It absorbs and registers every experience, however tiny, and compares it with past experiences, then stores it away. You can say, "Spaghetti again? We had it twice last week," because your brain stores information by constantly comparing today with yesterday. At the same time, you develop likes and dislikes, grow bored, long for variety, and reach the end of one phase of your life, ready for the next. The brain enables it all to take place. It constantly connects new information with what you learned in the past. You remodel and refine your neural network on a second-by-second basis, but so does the world you experience. The largest super computer in existence cannot match this feat, which all of us take for granted.

The brain isn't daunted by its endless tasks. The more you ask it to do, the more it can do. Your brain is capable of making a quadrillion (one million billion) synapses. Each is like a microscopic telephone, reaching any other telephone on the line as often as it wants. Biologist and Nobel laureate Gerald Edelman points out that the number of possible neural circuits in the brain are 10 followed by a million zeros. Consider that the number of particles in the known universe is estimated to be only 10 followed by seventy-nine zeros!

You may think you are reading this sentence right now, or looking out the window to check the weather, but actually you are not. What you are actually doing is outstripping the universe. That's a fact, not science fiction. Occasionally this fact intrudes into an everyday life with astonishing results. When it does, complexity is either a friend or enemy, and sometimes a little of both. One of the most exclusive clubs in the world consists of a handful of people who share a mysterious condition that was discovered only recently, in 2006: hyperthymesia. They remember everything. They have total recall. When they get together, they can play mental games like: What's the best April 4 you ever had? Each person rapidly flips through a mental Rolodex, but instead of note cards, they see the actual events of every April 4 in their lives. Within a minute someone will say, "Oh, 1983, definitely. I had a new yellow sundress, and my mother and I drank Orange Crush on the beach while my dad read the paper. That was in the afternoon; we went to a seafood restaurant for lobster at six."

They can recollect any day of their lives with complete, unerring accuracy. (*Thymesia*, one of the root words in *hyperthymesia*, is Greek for "remembering." The other word, *hyper*, means "excessive.") Researchers have located only seven or eight Americans to date who exhibit this condition, but it isn't a malady. None of these people have brain damage, and in some cases their ability to remember every detail of their lives began suddenly, on a specific day, when ordinary memory took a quantum leap.

To qualify for the diagnosis of hyperthymesia, a person has to pass memory tests that seem impossible. One woman was played the theme song from a sitcom that ran on television for only two episodes in the 1980s, but having seen one of them, she instantly knew the show's name. Another candidate was a baseball fan. She was asked to recall the score in a certain game between Pittsburgh and Cincinnati years before. "That's a trick question," she replied. "The team plane broke down, and Pittsburgh never arrived. The game was a forfeit."

We discussed memory in the previous chapter of this book, and hyperthymesia is the ultimate example of an ability that everyone shares being carried to superhuman lengths—only, it's very human still. When asked whether she liked having perfect recall, one subject sighed. "I can remember every time my mother told me I was too fat." Those with hyperthymesia agree that revisiting the past can be acutely painful. They avoid thinking about the worst experiences in their lives, which are unpleasant for anyone to recall but extraordinarily vivid for them, as vivid as actually living them. Much of the time their total recall is uncontrollable. The mere mention of a date causes a visual track to unspool in their mind's eye, running parallel to normal visual images. ("It's like a split screen; I'll be talking to someone and seeing something else," reports one subject.)

You and I don't have hyperthymesia, so how does it relate to the goal of super brain? The problem of complexity enters the picture. Science has studied total recall and the brain's memory centers; several are enlarged in people with hyperthymesia. The cause is unknown. Researchers suspect links to obsessive-compulsive disorder (OCD), since people with hyperthymesia often display compulsive behaviors; or to various forms of attention deficit, since total recallers cannot shut down the memories once they start flooding in. Perhaps these are people who never developed the ability to forget. One thing can always be counted on with the human brain: you can't look anywhere without looking everywhere.

Looking for Heroes

The way to get around the problem of complexity is to turn it on its head. If your brain is ahead of the universe, then its hidden potential must be far greater than anyone supposes. We can leave those quadrillion connections to the neuroscientists. Let's pick three areas where, in a normal healthy brain, peak performance is reachable. In each area there will be someone who has led the way. These are heroes of super brain, even though you may not have seen them that way before.

HERO #1

ALBERT EINSTEIN
FOR ADAPTABILITY

Our first hero is the great physicist Albert Einstein, but we are not choosing him for his intellect. Einstein—like geniuses in general—is a paragon of success. Such people are intelligent and creative far beyond the norm. If we knew their secret, each of us would have greater success, no matter what we pursued. Highly successful people don't merely have seven habits. They use their brain in a way that is keyed to success. If you shut yourself out from Einstein's way of using his brain, you limit your possibilities for success. It isn't a matter of just "good genes." Einstein used his brain in a way that any person can learn.

The key is *adaptability*.

Super brain takes advantage of your innate ability to adapt. This ability is necessary for survival. Of all living things, humans have adapted to all environments on the planet. Confront us with the harshest climate, the strangest diets, the worst diseases, or the most fearsome crises presented by natural forces, and we adapt. *Homo sapiens* does this so incredibly well that we take it for granted until someone appears before us who carries adaptability to a new level, someone like Einstein.

Einstein adapted by facing the unknown and conquering it. His field was physics, but the unknown confronts everyone on a daily basis. Life is full of unexpected challenges. To adapt to the

unknown, Einstein developed three strengths and avoided three obstacles:

Three strengths: Letting go, being flexible, hanging loose
Three obstacles: Habits, conditioning, stuckness

You can measure a person's adaptability by how much they are
able to let go, remain flexible, and hang loose in the face of difficulties. You can measure how poorly a person adapts by the dominance
of old habits and conditioning that keep them stuck. Harmful memories of shocks and setbacks in their past tell them over and over how
limited they are. Einstein was able to ignore old habits of thought
that surrounded him. He hung loose and let new solutions come
to him through dreaming and intuition. He learned everything he
could about a problem, then surrendered to unknown possibilities.

This isn't how the public views Einstein, who is imagined as a
brainiac with wild frowzy hair filling the blackboard with mathematical equations. But let's look at his career from a personal perspective. As he tells it, Einstein's great motivation was awe and
wonder before the mysteries of Nature. This was a spiritual state,
and he would say that penetrating the secrets of the universe was
like reading the mind of God. By seeing the cosmos first as a mystery, Einstein was rejecting the habit of seeing it as a giant machine
whose moving parts could be figured out and measured. That was
how Isaac Newton had viewed physics. Remarkably, Einstein took
the most basic notions in the Newtonian system, such as gravity and
space, and totally reinvented them.

He did so, as the whole world soon learned, through the theory
of relativity and his famous equation, $E=mc^2$. Higher mathematics
was involved, but that's a red herring. Einstein once told some young
students, "Do not worry about your problems with mathematics. I
assure you mine are far greater." This wasn't false modesty. His creative method was more like dreaming than cogitating; once he "saw"

how time and space worked, devising the mathematical proof came later, with much difficulty.

When you face a new problem, you can solve it in old ways or in a new way. The first is by far the easier path to follow. Think about an old married couple who argue all the time. They feel frustrated and blocked. Neither wants to give an inch. The result is a ritual, in which they repeat the same stubborn opinions, make the same nagging complaints, exhibit the same inability to accept the other's point of view. What would be a new way to get an old married couple out of their misery?

Instead of remaining stuck in the old behaviors, which are wired into their brains, they could use their brains in the following ways:

HOW TO BE ADAPTABLE

Stop repeating what never worked in the first place.

Stand back and ask for a new solution.

Stop struggling at the level of the problem—the answer never lies there.

Work on your own stuckness. Don't worry about the other person.

When the old stresses are triggered, walk away.

See righteous anger for what it really is—destructive anger dressed up to sound positive.

Rebuild the bonds that have become frayed.

Take on more of the burden than you think you deserve.

Stop attaching so much weight to being right. In the grand scheme of things, being right is insignificant compared with being happy.

Taking these steps isn't simply sound psychology: it creates a space so that your brain can change. Repetition glues old habits into the brain. Nursing a negative emotion is the surest way to block positive emotions. So every time an old married couple revisits the same resentments, they are wiring them harder into their brains. Ironically, Einstein, a master at applying such amazing adaptability to physics, saw himself as a failure as a husband and father. He divorced his first wife, Mileva, in 1919 after living apart for five years. A daughter born out of wedlock in 1902 has disappeared from the pages of history. One of his two sons was schizophrenic and died in a mental asylum; the other, who suffered as a child when his parents separated, was alienated from his father for two decades. These situations caused Einstein much pain. But even for a genius, emotions are more primitive and urgent than rational thoughts. Thoughts move like lightning; emotions move much more slowly and sometimes almost imperceptibly.

Here is a good place to point out that separating emotions and reason is totally artificial. The two are merged. Brain scans affirm that the limbic system, a part of the lower brain that plays a major role in emotions, lights up when people think they are making rational decisions. This is inescapable because the circuitry of the brain is entirely interconnected. Studies have shown that when people feel good, they are willing to pay unreasonable prices for things. (Pay three hundred dollars for jogging shoes? Why not, I feel great today!) But they are also willing to pay more when they feel depressed. (Six dollars for a chocolate chip cookie? Why not, it will cheer me up.) The point is that we make decisions against an emotional background, even if we rationalize that we don't.

Part of adaptability is to be aware of the emotional component instead of denying it. Otherwise, you run the risk that your brain will start using you. The economist Martin Shubik devised an unusual auction, in which the object put up for sale was a dollar bill. You might assume that the winning bid was $1, but it wasn't, because in this auction, the winner got the dollar bill, but whoever made the second-highest bid had to pay that amount to the auctioneer. Thus if I win by bidding $2 and you lose by bidding $1.50, you must hand over that amount, with nothing to show for it.

When this experiment was run, the bidding went well above a dollar. Typically, two male students were the last bidders standing. They felt competitive; each wanted to punish the other; neither wanted to be the loser who got punished. Whatever their motives, irrational factors sent the bidding higher and higher. (One wonders why it didn't skyrocket, ending only when one bidder ran out of money.)

Just as interesting is the fact that when experimenters try to eliminate the emotional side of decision making, they fail. No one has yet run a study where the subjects made purely rational decisions. We pay a high premium for stubbornly sticking to our opinions, backed by stuck emotions, habits, memories, and beliefs.

Bottom line: If you want to achieve success in any field, become like Einstein. Maximize your brain's ability to adapt.

YOU ARE BECOMING MORE ADAPTABLE WHEN

You can laugh at yourself.

You see that there's more to the situation than you realize.

Other people no longer look like antagonists simply because they disagree with you.

Negotiating starts to work, and you genuinely participate in it.

Compromise becomes a positive word.

You can hang loose in a state of relaxed alertness.

You see things in a way you didn't before, and this delights you.

HERO #2
A NEWBORN BABY
FOR INTEGRATION

Our next hero isn't famous or a genius or even gifted. It's every newborn baby. Babies are paragons of health and well-being. Every cell in their bodies is vibrantly alive. They see the world as a place of endless discovery. Each day, if not each minute, is like a new world. What makes for their state of robust well-being isn't that they are born in a good mood. Rather, their brains are constantly on the move, reshaping themselves as the world expands. Today is a new world, whether you are a baby or not, if it expands on what you experienced yesterday.

Babies have not shut themselves down or become stuck in old, outworn conditioning. Whatever their brains absorbed yesterday remains in place while new horizons keep opening up: walking, talking, learning how to relate and feel. When we grow up, we become nostalgic about the innocence of childhood. We sense a loss. What have we lost that babies have in abundance?

The key is *integration*.

Among all living things, human beings absorb every possible input and integrate it—that is, we make a whole picture. At this very minute, just like a newborn baby, you are sifting through billions of bits of raw data to form a coherent world. Here, *sift* is a technical term proposed by psychiatrist Daniel Siegel. It stands for

S—Sensation
I—Image
F—Feeling
T—Thought

Nothing is real except through these channels: either you sense it as a sensation (like pain or pleasure), imagine it visually, feel it emo-

tionally, or think about it. Sifting goes on constantly, and yet is ut-
terly mysterious. Imagine a beautiful sunset in your mind's eye. No
photons of light hit your retina, as they would if you were gazing
at an actual sunset. No illumination lights up your visual cortex,
which is submerged in the same blackness as the rest of the brain.
Yet microvolts of electricity pumping ions back and forth along your
neurons magically produce a picture full of light, not to mention
beauty and a cascade of associations with every other sunset you've
ever seen. (How the brain correlates this image through physical
means with your imagination is a central mystery in the mind-brain
connection.)

Integrating bits of raw data into pictures of reality is a process
that reaches right down to the cellular level, because anything the
brain does is communicated to the rest of the body. Quite literally,
when you feel depressed or have a bright idea or think you are in danger, your cells join in. Technically, what's at work is a feedback loop
that integrates mind, body, and the outside world in one process. Incoming data stimulates the nervous system. A response arises. The
report of this response is sent out to each cell, and in return the cells
say what they think about it.

Babies are perfect feedback machines. You can learn from them
what it means to integrate your own personal reality with greater
success. Just consciously do what nature designed into the infant
brain.

HOW TO INTEGRATE FEEDBACK

Remain open to as much input as possible.

Don't shut down the feedback loop with judgment, rigid beliefs, and prejudices.

Don't censor incoming data through denial.

Examine other points of view as if they were your own.

Take possession of everything in your life. Be self-sufficient.

Work on psychological blocks like shame and guilt—they falsely color your reality.

Free yourself emotionally—being emotionally resilient is the best defense against growing rigid.

Harbor no secrets—they create dark places in the psyche.

Be willing to redefine yourself every day.

Don't regret the past or fear the future. Both bring misery through self-doubt.

One way or another, you will inevitably create a reality around your own viewpoint. No one is perfect at integrating the world without bias. But babies teach us how to make our reality more complete. From birth, nature has designed us to approach the world as a whole, and when we slice experience up into bits and pieces, wholeness breaks down. Then, instead of living in reality, you are being fooled by a reality illusion.

Think of a dictator who has become used to unquestioned power. He stays in place through terror and a secret police. He bribes his enemies or makes them disappear in the middle of the night. Typically, such dictators are astounded when opposition rises up, and up to the moment when they are deposed or murdered by a mob, they believe that they are justified. They even fantasize that the people who suffer oppression in a police state love their oppressor. This is reality illusion carried to an extreme.

The fall of dictators fascinates us at another level because we sense, somewhere deep down, that unlimited power could do the same to us. A dark magic seems to draw a veil over the eyes of the deluded. But when it comes to the reality illusion that everyone lives in, there is no dark magic. There is only a failure to integrate. We are born with the ability to create wholeness, yet we choose denial, repression, forgetting, inattention, selective memory, personal bias, and old habits instead. These influences are hard to overcome. Inertia is on their side, for one thing. But you can't feel balanced, safe, happy, and in tune until you regain the wholeness that comes naturally to every newborn. This is the key to well-being as well as physical health.

Being a fully integrated person means having three strengths that reflect a baby's approach to the world and avoiding three obstacles that plague us as adults.

> **Three strengths**: Communicating, staying balanced,
>   seeing the big picture
> **Three obstacles**: Isolation, conflict, repression

When you are in an integrated state, either in body or mind, you communicate openly. You know what you feel; you express it; you absorb signals from everyone around you. But countless adults experience a breakdown in communication. They feel isolated from all kinds of things: their feelings, other people, the jobs they go to every morning. They become entangled in conflicts, and as a result, they learn to repress what they really feel and all their real desires. These feelings are not just psychological factors. They affect the brain and, in turn, every cell in the body.

Bottom line: If you want to return to the natural state of health and well-being, be like a newborn baby. Integrate your experiences into wholeness instead of living with separation and conflict.

YOU ARE BECOMING MORE INTEGRATED WHEN
> You create a safe place where you can be yourself.
> You invite others into the same safe space so that they can
>   be themselves.

You desire to know yourself.

You look at areas of denial, accept hard truths, and face reality.

You make peace with your dark side, using it as neither a secret ally nor a feared enemy.

You honestly assess and heal guilt and shame.

A sense of higher purpose dawns.

You feel inspired.

You offer yourself in service to others.

Higher reality seems real and attainable.

HERO #3
THE BUDDHA
FOR EXPANSION OF CONSCIOUSNESS

We use our brains first and foremost to be conscious, and some people take their consciousness much farther than others. Our heroes, our paragons for inner growth, are the spiritual guides of humanity wherever they arise. One particular hero, the Buddha, and the type he represents—saints, sages, and visionaries—display to perfection a unique trait of human beings: to live for meaning, which leads to a craving for the highest meaning. Meaning comes from within. It goes beyond the brute facts of life. The raw data that streams into the five senses is meaningless by itself. Looking at the brief, brutal lives of Paleolithic cave men or early hunters and gatherers, you would never suspect that their brains were capable of mathematics, philosophy, art, and higher reason. Those capacities were hidden, and a figure like the Buddha, who lived amid the poverty and struggle of life in India more than two thousand years ago, indicates that much more remains hidden inside us, if only we could tap into our yearning for meaning.

The key is *expansion of consciousness.*

No matter what kind of experience you are having, having it presupposes that you are aware. To be human is to be conscious—the only issue is how conscious. If you strip away all overtones of religion and mysticism, the state of higher consciousness that Buddha exemplifies is part of everyone's inheritance. An old Indian adage compares consciousness to a lamp at the door, shining into the house and out into the world at the same time. It makes you aware of things "out there" and "in here" simultaneously. Being aware creates a relationship between the two.

Is that relationship good or bad? The heavens and hells conceived in the human mind are all products of thought. We think our way into them, and we think our way out. "You are only as safe as your thoughts," says a wise aphorism. But where do thoughts come from—the dangerous, unsafe kind as well as the reassuring, trusting kind? They originate in the invisible realm of consciousness. For the mind, awareness is the womb of creation. To achieve a life filled with meaning, you must figure out how to be more conscious; only then do you become the author of your own destiny.

HOW TO EXPAND YOUR CONSCIOUSNESS

Put a higher value on being awake, aware, and alert.

Resist conformity. Don't think and act like everyone else.

Value yourself. Don't wait for the approval of others to validate you. Instead of desiring external validation, strive to help others.

Expose your mind to a higher vision through art, poetry, and music. Read widely in world scriptures and holy texts.

Question your own core beliefs.

Work on reducing the demands of the ego. Expand beyond the limits of "I, me, mine."

Aim for the highest meaning that your life can have.

Keep faith that inner growth is an unending process.

Walk the spiritual path, however you define it, with sincerity and hope.

Consciousness is a curious thing; we all have it, but we never have enough. Yet the supply is infinite. Because he stands for this eternal unfoldment, the Buddha is more than Buddhism. The greatest spiritual guides exemplify three strengths and avoid three obstacles.

Three strengths: Evolving, expanding, being inspired
Three obstacles: Contraction, fixed boundaries, conformity

None of these terms are overtly religious. They are about facing existence with more awareness. According to legend, when the Buddha was a troubled seeker named Siddhartha, the king, his father, wanted his son, born a prince, to grow into a great ruler. To suppress Siddhartha's spiritual yearnings, his father kept him imprisoned behind the palace walls, surrounded him with luxury, and refused to allow him any contact with the suffering of everyday life. That is a parable for what we do to our own awareness. We contract behind the walls of ego. We refuse to look beyond fixed mental boundaries. We pursue the pleasures and possessions that a consumer society holds out for us.

Higher consciousness isn't necessarily a spiritual state—it is an expanded state. Spirituality arrives in due course, depending on how contracted you were when you started. A life filled with stress and

sorrow naturally causes awareness to contract; it is a survival response, like a herd of antelope drawing together at the approach of
a lion. You need to realize that contraction may create a primitive
kind of safety, but at the cost of tightness, fear, constant vigilance,
and insecurity. Only by expanding your consciousness can you be a
lamp at the door, seeing the world without fear and yourself without
insecurity.

Bottom line: If you want to achieve inner growth, be more like
the Buddha in your approach to consciousness. Expand your awareness and look beyond the walls set up in the mind.

YOU ARE BECOMING MORE CONSCIOUS WHEN

You can speak your own truth.

You no longer see good and evil as fixed opposites. Gray
areas emerge, and you accept them.

You forgive more easily because you understand where other
people are coming from.

You feel safer in the world. You see that the world is as you
are.

You feel less isolated and alone, which shows that you are
basing your happiness on yourself, not on others.

Fear is no longer as persuasive as it used to be.

You see reality is a rich field of possibilities, and you are
eager to explore it.

You escape the clutches of "us versus them" thinking in
religion, politics, and social status.

You do not feel threatened by or fear the unknown. The
future is born in the unknown and nowhere else.

You can see wisdom in uncertainty. This attitude allows life
to flow naturally, without the need to make things black
or white.

You see being here as its own reward.

Heroes of super brain aren't the same as super heroes. They are realistic models for change. We believe that the continual development of super brain will lead to a healthier and more highly functioning brain. You will allow your emotions and thoughts to serve their intended purpose, to create the reality in which you deeply desire to live. You will no longer identify with repetitive cyclic patterns in the brain or the limited behavior that they lead to. You will be free to experience higher awareness and a more powerful sense of who you really can be.

SUPER BRAIN SOLUTIONS

DEPRESSION

In this chapter, we've taken another step toward showing you how to use your brain instead of letting it use you. Applying this principle to depression, which afflicts millions of people—it is the leading disability among Americans between fifteen and forty-five—will result in a world of good. There is no more painful example of people being used by their brain. As one former patient described it, "I felt as if I was falling just before I hit the floor, only instead of lasting for a second, the panicky feeling lasted for days and days, and I didn't even know what I was afraid of." Sufferers from depression feel victimized by a brain gone awry.

Even though depression is classified as a mood disorder, traceable to the brain's inability to react properly to inner and outer stress, it affects the whole body. It throws off bodily rhythms, in terms of sleep irregularity. It causes a loss of interest in sex and decreases appetite. Depressed people meet eating and making love with fatigued indifference. In social situations they feel disconnected. They cannot clearly understand what other people are saying to them; they can't express to others how they feel—being with others is a disturbing blur.

The brain is involved in all these full-body symptoms. The brain scans of depressed people show a unique pattern in which some areas of the brain are overactive and others underactive. Depression typically affects the anterior cingulate cortex (involved in negative emotions, but also empathy), the amygdala (responsible for emotions and for responding to novel situations—depressed people generally don't respond well to new things), and the hypothalamus (involved in drives like sex and appetite). These interconnected areas link into a kind of depression circuit—the network we want to affect positively in order to return to normal.

Depression is caused by a trigger, but the trigger can be so small that it passes unnoticed. Once it has been triggered the first time, the brain changes, and then in the future it takes smaller and smaller triggers to enter depression, until finally almost none is needed. When that happens, the person becomes a prisoner of the runaway emotions that can lead to mood disorders.

Are you depressed? We all use the word casually, but being sad or down isn't the same as being depressed. To be diagnosed with depression, either acute (short term) or chronic (long term), your moods stop the normal pattern of swinging back and forth. You cannot shake a feeling of sadness, helplessness, and hopelessness, or become interested in things around you. Everyday activities feel overwhelming. Freud connected depression to grief, and the two conditions are similar. In many cases, just as grief naturally lifts after a time, so can depression. But if it lingers, the person faces every day without hope of relief. The person sees his life as a total failure and may see no reason to keep living. (Around 80 percent of suicides are caused by a bout of major depression.)

People who have long-term depression often cannot pinpoint when their symptoms started, or why. They may feel that the key is genetic if depression runs in their family, or they may have a loose recollection of when they first noticed that they were sad all the time or felt hopeless for no apparent reason. Depression, along with autism, is considered the most genetic of psychological disorders; up to 80 percent of sufferers have someone else in the family who is or was depressed. But in most cases, genes merely predispose a person to mood disorders while not ensuring onset. To bring on a psychiatric illness, genes and environment work in concert.

Many depressed people will tell you that their problem is not the feeling of depression itself but the overwhelming fatigue they experience—as someone said, the opposite of being depressed isn't being happy, it's being vital. Fatigue, in turn, leads to more depression. Once you decide with conscious awareness and unwavering intent that you are not your brain, you can be one with your emotions and reactions to the outside world. Acting as the leader of your brain, you can actively reprogram your own neurochemistry and even genetic activity, no longer indentured to mood disorders.

The key is to get the stuck or imbalanced parts of your brain to move again. Once that happens, you can go on over time to bring the brain back into natural balance. That's the goal we'd like to help advance, and it's also the most holistic approach.

Three Steps in Depression

Once the brain has been trained, its responses feel normal. Sometimes depressed people have adapted so well that they are surprised when a friend, doctor, or therapist tells them that they are depressed. Various theories about genetic influence and chemical imbalances in the brains of depressed people are still widespread, but such explanations have fallen under a shadow of doubt. (Basic research has revealed that depressed patients don't differ genetically from others. Nor is it clear that antidepressants work by correcting a chemical imbalance. But when depressed patients get the right psychotherapy, talking through their feelings, their brains change in a way that resembles the changes produced by drugs. So another mystery is added: how can talking and taking a pill produce the same physiological result? No one knows.) If you met a young person with bad table manners, what would you ascribe it to? Most likely you would suppose that this behavior started somewhere in childhood and turned into a habit. If the habit persisted, it's because the person saw no good reason to change it. What if depression shares the same profile? We could retrace the steps of how depression developed and then undo them.

So let's look at depression as a fixed behavior. Fixed behaviors have three components:

1. An early outside cause, often since forgotten.
2. A response to that cause, which for some reason is unhealthy or unexamined.
3. A longstanding habit that becomes automatic.

Let's rid our minds of calling every kind of depression a disease, particularly the mild to moderate depression that most sufferers experience. (Certainly severe, chronic depression should be approached like other severe mental disorders.) If you get depressed after a bad divorce, you're not sick. If you grieve over a loss or feel down after losing your job, it's not illness. When someone loses a beloved spouse, we may say, "She's out of her mind with grief," but grief is natural, and the depression that comes with it is also natural. What this tells us is that depression is a natural response that can go terribly wrong.

When depression goes wrong, all three components are to blame:

1. *Outside causes*: Outside events can make anyone depressed. During the severe 2008 economic recession, 60 percent of people who lost their job say it made them anxious or depressed. The number is much higher among workers who had been laid off for more than a year. If you subject yourself to enough stress over a long period of time, depression is much more likely. Long-term stress can be caused by a boring job, a sour relationship, prolonged stretches of loneliness and social isolation, and chronic disease. To some extent, a depressed person is reacting to bad circumstances, either now or in the past.
2. *The response*: An outside cause cannot make you depressed unless you respond in a certain way. People who are depressed learned long ago to have a skewed response, such as the following, when something went wrong in their lives:

It's my fault.
I'm not good enough.

Nothing will work out.
I knew things would go wrong.
I can't do anything about it.
It was just a matter of time.

Young children who have any of these responses feel they make sense. They are reporting to their brains with a viewpoint on reality. The brain conforms to the picture of reality it is trained to see. Small children have little control over their lives; they are weak and vulnerable. An unloving parent can create any of these responses, and so can a disastrous family event like a death. But when adults have these responses, then the past is undermining the present.

3. *The habit of being depressed*: Once you have a depressed response, it reinforces the next response when you face a new stress from the outside world. Did your first boyfriend dump you? Then it's natural to fear that the second one might, also. Some people can manage this fear, but for others it looms large. Instead of daring to find a second boyfriend who is more loving and loyal, they turn the blame and fear inward. They keep having depressed responses, generated from the inside, and after a while these responses turn into a habit.

Undoing the Past

Once a person's depression turns into a habit, which probably happens years earlier than their recognition of being sad and hopeless, it no longer needs an outside trigger. Depressed people are depressed about being depressed. A gray film coats everything; optimism is impossible. This defeated state tells us that the brain has formed fixed pathways, and that perhaps—or probably—genetic output and neurotransmitters are involved. The person's whole support system for creating his or her personal reality comes into play.

When the depressed response is internalized, it's like smoldering hot coals that will flame up with only a small stir. A minor incident like a flat tire or a bounced check leaves the person no room for deciding "Is this going to bother me or not?" The depressed response is already wired in. Depressed people can even feel sad about good news; they are always waiting for the other shoe to drop, because they are trapped in the habit of depression. The brain's imbalance can be traced to mental activity. Brain scans of depressed people appear to support this connection. They show that the same areas that light up due to the beneficial effects of antidepressants also light up if the person enters therapy and successfully talks through their depression. Talk is a form of behavior.

If behavior can get you out of depression, it's only reasonable to suppose that behavior can get you into it. (For the moment we'll set aside the kind of depression that has physical—or as doctors term it, organic—causes, such as many diseases and senile dementia as well as poor diet and environmental toxins. When the physical cause is corrected, the depression usually goes away automatically.) Since this explanation sounds reasonable, the key questions are how to avoid getting into the depressed response and how to reverse depression once it sets in. We can approach the issues of prevention and getting better using the same three categories that we have been discussing.

Outside events: People will say, "Did you see the evening news? I'm so depressed about the state of the world." Or "I was depressed during the whole period around 9/11." Outside events can make us depressed, but in fact they are the least powerful ingredient in causing depression. Losing your job can be depressing if you are prone to the depressed response, but if you are not, it can spur you to rise even higher. Bad things are unavoidable, but some factors make them worse:

The stress is repeated.
The stress is unpredictable.
You have no control over the stress.

Consider a wife whose husband is an abusive rage-aholic. He has hit her repeated times; she cannot predict when he will fly into one of his rages; she cannot find the will or the strength to leave him. Such a woman will be a strong candidate for depression because all three elements of a major stress are present. The abuse being directed at her is repeated, unpredictable, and beyond her control.

Her whole mind-body system will start to shut down if she remains in this situation. That happens when mice are given mild electric shocks. When researchers space the shocks at random intervals, give them over and over, and provide no way for the mice to escape, it doesn't matter that the shocks are harmless. The mice will soon give up, act lethargic and helpless, and in time die. In other words, their induced depression is so extreme that it has destroyed the will to live.

What does this mean for you, the person who wants to avoid depression? First, stop exposing yourself to stresses that occur over and over. This could mean a bad boss, an abusive husband, or any other stress that is reinforced every day. Second, avoid unpredictability of the stressful kind. Yes, life is uncertain, but there's a limit to what uncertainties are acceptable. A boss who unpredictably flies into a rage isn't acceptable. For many people, a sales job, where any customer might lash out or slam the door in your face, is too uncertain to bear. A spouse who may or may not cheat is unpredictable in the wrong way.

By the same token, you should increase the predictable routines that help defend against stress. Everyone needs a good night's sleep, regular exercise, a steady relationship, and a job they can count on. The regular habits aren't just good for you in a vague general way— they help you avoid depression by training your brain in a positive direction.

As part of feeling helpless and hopeless, depressed people tend to be passive in stressful situations. Unable to see a fruitful way to fix such a situation, they deny themselves key decisions that might work; instead they lean toward making no decision, which rarely works.

They put up with the bad situation for too long. When depression isn't present, you can generally figure out what to fix, what to put up with, and what to walk away from. Those are basic choices that you must make throughout your life.

If you know that you are prone to depression, it's important for you to deal with problems more promptly and directly than you otherwise might, because the longer you wait, the more chance you give the depressed response to set in. I'm speaking of ordinary situations like a potential conflict at work, a teenager at home who is overstepping his curfew, or a partner who isn't doing his part of the housework. Depression makes you overly sensitive to small triggers, leading to a sense of helpless resignation. But if you act early, before you reach this stage, you have room to manage an everyday stress and the energy to carry out your decision to do so. Learn how to make such decisions promptly, ignoring the little voice that warns you not to make waves. You're not making waves; you're heading the depressed response off at the pass.

The depressed response: Subtler causes of depression are more difficult to undo than outside stress. If you don't want to be overweight, it's much easier to avoid putting on the pounds in the first place than losing them once they are on. The same holds true for depression. It's much easier to learn the right response to stress than to unravel the wrong one. The right response involves emotional resilience, which allows you to let go of stress rather than to take it in. Undoing the wrong response requires retraining your brain. But even so, some overweight people manage to lose the pounds, and a brain that has been trained to respond with depression can be untrained.

We all have self-defeating responses, and we don't like what they do to us. It takes time and effort to replace them with better alternatives. In the case of depression, it is now well recognized that changing the self-defeating beliefs of a depressed person can lead to recovery. Beliefs are like software programs that keep repeating the same commands, only beliefs are more pernicious, since they dig in deeper with every repetition.

Here are some examples of the ingrained programming that automatically comes into play when you're feeling depressed, followed by alternative beliefs that counter the depressed response:

TRADING OUT TOXIC BELIEFS

1. It's all my fault.
Instead, you could think: It's not my fault, it's nobody's fault, the fault hasn't been determined yet, it may be nobody's fault, *or* finding fault does no good—we should be focusing on the solution.

2. I'm not good enough.
Instead, you could think: I am good enough, I don't need to compare myself to others, it's not about good or bad, "good enough" is relative, I'll be better tomorrow, *or* I'm on a learning curve.

3. Nothing will work out.
Instead, you could think: Something will occur to me, things have a way of working out, I can ask for help, if one thing doesn't work out, there's always something else, *or* being pessimistic doesn't help me find a solution.

4. I knew things would go wrong.
Instead, you could think: No, I didn't know, I'm second-guessing, I'm just feeling anxious, it will pass, *or* looking backward is only good if it leads to a better future.

5. I can't do anything about it.
Instead, you could think: I can do something about it, I can find someone to do something about it, I always have the option of walking out,

I need to study the situation more thoroughly, *or* being defeatist isn't helping me make things better.

6. It was just a matter of time.

Instead, you could think: I'm not a fatalist, this was unpredictable, this too shall pass, it never rains all the time, *or* being fatalistic robs me of free choice.

We are not saying that all the alternative beliefs work all the time. You must be flexible. The nasty trick of the depressed response is that it paints everything with the same brush. You feel helpless about repairing your car's transmission (who wouldn't?) but also about getting out of bed to face the day (a sign of depression). To become flexible, you must beat the depressed response at its own game.

How to do that? If your automatic reaction is associated with sadness, helplessness, and hopelessness, refuse to accept it. Give yourself a moment, take a deep breath, and refer to our list of alternative responses. Find one that works. This takes time and effort, but it will pay off. Learning a new response forms new neural pathways in the brain. It also opens doors. What kinds of doors? When you are depressed, you tend to be isolated, lonely, apathetic, inactive, passive, and closed to change. The new doors have exactly the opposite effect. By introducing a new response, you resist the temptation to fall back on old, stale beliefs. Instead of being isolated, you realize that other people are good for you. Instead of being passive, you see that taking charge is also good for you.

Another strategy is to break down the depressed response, which feels so overwhelming, into manageable pieces. The best tactic is to take one step at a time, choosing a piece you feel ready to handle. Inertia is depression's best friend. You will always have a hump to get over before you can actually do something positive. So don't turn the hump into a Himalayan peak.

Pushing yourself over the smallest hump urges the brain to give up an old pattern for a new one. You are actually expanding your awareness when you let in fresh impulses from the source, which is the real you. Behind the mask of depression, which is a behavior tied to a fixed response, lies the real you, the core self that can direct the healing process. To put it simply, you alone have the power to create healing. Depression creates the illusion that all your power has been stripped away. In truth, once you find an opening, you can reclaim the real you, step by step.

The habit of depression: If you have ever lived around an alcoholic or any other addict, you know that they behave in predictable pendulum swings. When they are sober or off the drug, they sincerely repent and never want to return to their habit. But when the addict is faced with a temptation to drink or shoot up or overeat or fly into a rage (depending on what their habit happens to be), their good intentions fly out the window. Willpower disappears, the habit takes control, and only getting a fix matters.

Depression also has an addictive side, in which sadness and hopelessness take charge. "I can't be any other way" is the common cry of both the addict and the habitually depressed person. In many cases, a "good me" and a "bad me" are warring against each other. For the alcoholic, the "bad me" drinks, while the "good me" is sober. For the depressed person, the "bad me" is sad and hopeless, while the "good me" is happy and optimistic. But in truth depression casts its shadow over everything. The best moments are merely a prelude to a relapse. The "bad me" is going to win in the end; the "good me" is merely its pawn.

The war is unwinnable, every victory is only temporary, and the pendulum keeps swinging back and forth. When a war is unwinnable, why fight? The secret to beating any fixed habit is to stop fighting with yourself, to find a place inside that isn't at war. In spiritual terms, that place is the true self. Meditation opens the way to reaching it; the world's wisdom traditions affirm that no one can be denied peace, calm, silence, the fullness of joy, and reverence for life. When people frown and tell me that they don't believe in meditation, my response is that they must not believe in the brain, because four decades of brain research have proven that the brain is transformed by meditation, and now newer evidence suggests that genetic output also improves with meditation. That is, the right genes get switched on and the wrong ones switched off.

To challenge the depressed response, it's not sufficient to simply go inward. You must activate your real self and bring it into the world. Until you can prove the usefulness of new responses and beliefs, the old ones will keep a foothold in your consciousness. You are very used to them, and they know the quickest way to return. Therefore, breaking the habit of depression involves doing a combination of inner work and outer work, as follows:

WORKING BOTH SIDES

INNER WORK: CHANGING WHAT YOU THINK AND FEEL

Meditate.
Examine your negative beliefs.
Reject self-defeating reactions to life's challenges.
Learn new responses that are life-enhancing.
Adopt a higher vision of your life and live by it.
Recognize self-judgment and reject it.
Stop believing that fear is okay just because it's powerful.
Don't mistake moods for reality.

OUTER WORK: CHANGING YOUR BEHAVIOR

Reduce stressful conditions.
Find fulfilling work.
Don't associate with people who increase your depression.
Find people who are close to who you want to be.

Learn to give of yourself. Be generous of spirit.
Adopt good sleep habits, and exercise lightly once a day.
Focus on relationships instead of distractions and endless
consumerism.
Learn to re-parent yourself by finding mature, emotionally
healthy people who can love, who are accepting, and who
do not pass judgment.

Every physician and therapist has met hundreds of depressed people who desperately want help, but how many were on the road to recovery? Most put their faith in a pill or lapsed into a state of exhausted resignation. In some cases, drugs can relieve symptoms, but mild to moderate depression doesn't require a disease model, which often does no good. Current findings bear this out: in cases of mild to moderate depression, antidepressants barely surpass the placebo response (which leads to improvement on average in 30 percent of patients). They become more effective only when the depression grows more severe.

The three elements we have been focusing on—outside causes, the depressed response, and the habit of depression—offer a new approach. They give you the power to reverse the underlying conditions of your depression. We aren't saying that the cause of depression has been found, because in the end your depression is entangled with everything else in your life, including everything that is going on in your body.

Because of that, you must reshape your life on many levels, which you can only do consciously. Sometimes it takes very little to get out of depression, if escaping a bad job or a toxic marriage can be seen as simple. At least it's direct. At other times depression is like a fog that cannot be grabbed in any one place. But fogs can lift. The best news is that the real you isn't depressed and never has been. By setting out on the path to finding the real you, you will accomplish more than healing your depression. You will emerge into the light and see life in a new way.

PART 2

MAKING
REALITY

YOUR BRAIN,
YOUR WORLD

As you move through this book, you will see that mind, brain, and body work seamlessly together. Life is a continuous process. The deeper you master this process, the closer you are to arriving at the goal of super brain. A researcher like Rudy, looking at data about neuroplasticity, can marvel at how the brain creates new pathways. But the greater wonder is that mind can create matter. For that is actually happening in the brain, and it takes place thousands of times a second. Whether it is the rush you feel after winning the lottery, or the "fine free careless rapture" that the poet Robert Browning felt in the song of a thrush, both experiences require the brain to find a physical representation. Rapture needs chemistry, as does every other thought, feeling, and sensation. Neuroscience has established this fact quite firmly.

We want to take you to where real mastery lies, where "brain" doesn't sit in its earthbound compartment while "mind" floats airily above. The difference between them is man-made and misleading. Mind and brain are merged, and the place where super brain is born lies at the control switch that you can learn to operate.

The subtle regions of awareness are where the real power lies.

When someone steps up to receive the Oscar for best picture, they often exclaim, "This is a dream come true!" Dreams are subtle but powerful. Your personal vision sets the course of your life in motion. But first it must set the brain in motion, after which come action, possibilities, opportunities, lucky breaks, and everything needed to make a dream come true. This process we will call reality making. It's a continuous unfolding, and although science pays attention to the products of the brain—synapses, electrical potentials, and neurochemicals—these are gross expressions. Reality begins at a much subtler, invisible level.

How, then, do you take control of reality making? Some rules of the game apply, as follows:

THE RULES OF REALITY MAKING

You are not your brain.

You create everything about how the world looks and feels.

Perception isn't passive. You are not simply receiving a fixed, given reality. You are shaping it.

Self-awareness changes perception.

The more aware you are, the more power you have over reality.

Awareness contains the power to transform your world.

At a subtle level, your mind is merged with the creative forces of the universe.

We will explain the rules as we go. But reality making is natural and effortless, while at the same time it is almost beyond belief. The universe goes to the same place to create a star as you go to see a rose in your mind's eye. Now it's up to us to show why that incredible statement is true.

You Are Not Your Brain

The first principle in reality making is that you are not your brain. We already saw how crucial this insight is for people suffering from depression (as it is for people suffering from any other mood disorder, like anxiety, which is just as epidemic as depression). When you come down with a bad cold, no matter how much you are suffering, you don't say, "I am a cold." You say, "I have a cold." But linguistics is such that you don't say, "I have depression." You say "I am depressed," which means you identify with that condition. For countless people who are depressed and anxious, "I am" becomes extremely powerful. Mood colors the world. When you identify with being depressed, the world reflects how you feel. When you see a lemon, you don't think you are yellow, and the same should hold true with depression. In both cases, the mind is using the brain to create yellow, just as it is creating depression. There's an intimate link at the level of physiology, and if you control the link, you can change anything.

If the brain were in charge of your identity, it would make just as much sense to say "I am a yellow lemon" as it does to say "I am depressed." How, then, do we know the difference? How is it that you know that you are not a yellow lemon, while a depressed person may identify with his disorder so acutely that he commits suicide? Partly the difference is emotional. Biology comes into play here. The hippocampus is intimately wired to the amygdala, which regulates emotional memories and the fear response. In imaging studies when human subjects were shown a scary face while undergoing an fMRI (the best scan for showing brain activity in real time), the amygdala lit up like a Christmas tree. The fear response pours into the higher brain, which takes a while to realize that scary pictures are no reason to be afraid. Uncontrolled fears, even when they have no realistic cause, can lead to chronic anxiety and depression.

Biological counters can offset this effect. Recent studies suggest that new nerve cells in the hippocampus are able to inhibit the negative emotions evoked in the amygdala. Stress-alleviating activities, such as doing physical exercise and learning new things, promote the birth of new nerve cells, which, as we've seen, promotes neuroplasticity—new synapses and neural circuits. Neuroplasticity can directly regulate mood and prevent depression. Thus the birth of new nerve cells in the adult hippocampus helps overcome neurochemical imbalances that lead to mood disorders like depression.

In neuroscience this idea is novel, but in real life many people have discovered that going for a jog can lift them out of a blue mood. Because a yellow lemon doesn't trigger emotional responses while depression does, we've found an important difference at the level of the brain. Some studies have shown that antidepressants like Prozac may work at least in part by increasing neurogenesis (new nerve cells) in the hippocampus. In support of this idea, mice that are given antidepressants show positive changes in behavior, changes that can be preempted by deliberately blocking neurogenesis in the hippocampus.

The alert reader will point out that we seem to be arguing against ourselves. If Prozac makes you feel better, then what's wrong with taking a pill to promote desirable effects in the brain? First of all, drugs don't cure mood disorders—they only alleviate them. Once a patient stops taking an antidepressant or tranquilizer, the underlying disorder returns. Second, all drugs have side effects. Third, the beneficial effects of drugs wear out over time, requiring higher dosages to reach the same benefit. (In time there may be no benefit to reach.) Finally, studies have shown that antidepressants aren't as effective as their makers claim, and in most common cases of depression, couch therapy can achieve the same benefits. Our culture is addicted to popping pills as silver bullets, but the reality is that talking your way out of your depression is curative, whereas drugs by and large are not.

As the brain shifts, reality follows. Depressed people live not just in a sad mood but in a sad world. Sunshine is tinged with gray; colors lack luminosity. But those who have no mood disorder imbue the world with livelier qualities. A stoplight is red because the brain makes it red, although red-green color-blind people see the same stoplight as gray. Sugar is sweet because the brain makes it sweet, but for those who have lost their taste buds through injury or disease, sugar has no sweetness. Subtler qualities are at work, too. You add emotion to sugar's taste if it reminds you that you may be prediabetic; you add emotion to a stoplight if the sight evokes bad memories of a car accident in your past. The personal cannot be separated from the "facts" of daily life. Facts are personal, in fact. The radical part is that nothing escapes the process of reality making.

Every quality in the outside world exists because you create it. Your brain is not the creator but a translational tool. The real creator is mind.

It will take more to convince you that you are creating all of reality. We understand. Doubt arises from a widespread lack of knowledge about how the mind interacts with the world "out there."

Everything depends on the nervous system that is having the experience. Since humans don't have wings, we have no idea of a hummingbird's experience. Looking out an airplane window isn't the same thing as flying. A bird swoops and dives, balances in midair, keeps an eye out in all directions, and so on. A hummingbird's brain coordinates a wing speed of up to eighty beats per second and a heart rate of more than a thousand beats per minute. Humans cannot penetrate such an experience—in essence, a hummingbird is a vibrating gyroscope balanced in the middle of a whirling tornado of wings. You only have to consult a table of bird world records to be astonished. The smallest bird, the bee hummingbird of Cuba, weighs 1.8 grams, just over half the weight of a penny. Yet it has the same basic physiology as the world's largest bird, the African ostrich, which weighs around 350 pounds.

In order to explore reality, the nervous system must keep up with the new experience, monitor it, and control the rest of the body. The nervous systems of birds explore experience on the far horizon of flying. Water birds, for example, are designed to dive. Emperor penguins have been measured to dive to a depth of 1,584 feet. The fastest dive ever measured belongs to peregrine falcons studied in Germany—depending on the angle they took, the falcons reached a speed between 160 and 215 miles per hour. Birds' physical structure has adapted to push these boundaries. Their nervous systems are the key, not their wings or hearts. Thus a bird's brain has created the reality of flight.

This argument can be taken much further with the human brain, because our minds have free will, while a bird's awareness (so far as we can enter it) operates purely by instinct. For humans, a huge leap in reality making is possible.

But first, a note about something that Deepak is especially passionate about. It isn't correct to say that the brain "creates" a thought, an experience, or a perception, just as it isn't correct to say that a radio creates Mozart. The brain's role, like the transistors in a radio, is to provide a physical structure for delivering thought, as a radio allows you to hear music. When you see a rose, smell its luxurious scent, and stroke its velvety petals, all kinds of correlations happen in your brain. They are visible on an fMRI as they occur. But your brain isn't seeing, smelling, or touching the rose. Those are experiences, and only you can have an experience. This fact is essential; it makes you more than your brain.

To show the difference: in the 1930s, a pioneering brain surgeon named Wilder Penfield stimulated the area of the brain known as the motor cortex. He found that applying a tiny electrical charge to the motor cortex caused muscles to move. (Later research expanded on this finding extensively. Charges applied to memory centers can make people see vivid memories; doing the same to emotional centers can trigger spontaneous outbursts of feeling.) Penfield realized, however, that the distinction between mind and brain was crucial. Because brain tissue cannot feel physical pain, open-brain surgery can be performed with the patient awake.

Penfield would stimulate a local area of the motor cortex, causing the patient's arm to fly up. When he asked what had happened, the patient would say, "My arm moved." Then Penfield would ask the patient to raise an arm. When he asked what happened, the patient would say, "I moved my arm." In this simple direct way, Penfield showed something that everyone is aware of instinctively. There's a huge difference between having your arm move and moving it yourself. The difference lays bare the mysterious gap between mind and brain. Wanting to move your arm is an action of the mind; involuntary movement is an action triggered in the brain—they are not the same.

The distinction may sound finicky, but in the end it will be hugely important. For now, just remember that you are not your brain. The mind that gives orders to the brain is the only true creator, just as Mozart is the true creator of the music played on a radio. Instead of passively accepting anything in the world "out there," first claim your role as creator, which is active. Here is the true beginning of learning to make reality.

Creativity is based on making things new. Pablo Picasso often placed two eyes on the same side of a face, which bears no resemblance to nature (unless we are talking about flatfish like flounders and halibut, whose tiny fry are born with eyes on either side of their heads, only to have both eyes migrate to one side as they mature). Some people would accuse Picasso of making a mistake. There's a joke about a first-grade teacher taking her class to a modern art museum. Standing in front of an abstract painting, she says, "That's supposed to be a horse." From the back of the group a little boy pipes up, "Then why isn't it?"

But abstract painting makes "mistakes" in order to create something new. Picasso is seeing the human face in a new way. Because perception is endlessly adaptable, if you give Picasso a chance, you allow your own seeing to be distorted, compared to the ordinary way of looking at faces. A disturbed emotion arises, and all at once,

you may laugh or tremble nervously or find his abstract style beautiful. The new way excites you; you become part of it. The brain is
designed to allow everyone to make things new. If the brain were a
computer, it would store information, sort it in different ways, and
make lightning-fast calculations.

Creativity goes beyond that. It turns the raw material of life into
an entirely new picture, one never seen before. If you have Ham-
burger Helper for dinner five nights in a row, you can get bored,
complain, and wonder why life doesn't change. Or you can make
something new. Right now you are assembling your world like a
jigsaw puzzle in which every piece is under your control.

MAKING IT NEW
HOW TO TRANSFORM YOUR PERCEPTIONS

Take responsibility for your own experience.
Be skeptical of fixed reactions, both yours and anyone else's.
Confront old conditioning. It leads to unconscious behavior.
Be aware of your emotions and where they come from.
Examine your core beliefs. Hold them up to the light, and
    discard beliefs that make you stuck.
Ask yourself what part of reality you are rejecting. Freely
    consult the viewpoint of the people around you. Respect
    what they see in the situation.
Practice empathy so that you can experience the world
    through someone else's eyes.

These points all center on self-awareness. When you do
anything—eat breakfast, make love, think about the universe, write
a pop song—your mind can be in only one of three states: unconscious, aware, and self-aware. When you are unconscious, your

mind is passively receiving the constant stream of input from the outside world, with minimal reactions and no creativity. When you are aware, you pay attention to this stream of input. You select, decide, sort, process, and so on, making choices about what to accept and what to reject. When you are self-aware, you loop back on what you are doing, asking *How is this for me?* At any given moment, all three states coexist. We have no idea if that's true for a creature like the hummingbird. As its heart races at over a thousand beats per minute, is the bird thinking, *I'm tired?* That question comes out of self-awareness. Is it thinking, *My heart beats really, really fast?* That's a statement of simple awareness. We suppose, without knowing the truth, that a hummingbird isn't self-aware, and it may not even be aware. Its entire life could be spent unconsciously.

Unconscious, Aware, Self-aware

Human beings exist in all three states, and which one predominates at any given moment is up to you. Super brain depends on reducing our unconscious moments while increasing both awareness and self-awareness. Consider the fourth item in the previous list: *Be aware of your emotions and where they come from.* The first part is about awareness, the second about self-awareness. *I am angry* is an aware thought, while flying off the handle is unconscious. That's why we give latitude to someone who goes off into a rage, for example, at the scene of a car accident. We don't take what they say seriously until their rage is over and they calm down. Some legal systems forgive unconsciousness, allowing leniency for so-called crimes of passion. If you find your wife in bed with another man and react by strangling him on the spot, you are acting unconsciously, without full awareness.

It's good to be aware, but self-awareness is even better. *I am angry* gets you only so far if your aim is to control your anger. Knowing where your anger comes from adds the component of self-awareness. It allows you to see a pattern in your behavior. It takes into account that past outbursts haven't worked out so well. Maybe a spouse has left you in the past, or someone called the police. Once you bring in self-awareness, reality shifts. You start to take control; the power to change is dawning.

Awareness irrefutably penetrates the animal world. Elephants gather around a baby elephant that has died. They linger there and even return to sites of past deaths a year later. They huddle close to the mother who has lost her calf. If empathy means anything outside our human definition of it, elephants appear to empathize with one another. For all we know, a tiny hummingbird migrating thousands of miles from Mexico to Minnesota may be aware of what route it is taking, including visual signposts, the movement of the stars, and even the earth's magnetic field.

But we ascribe self-awareness only to ourselves. (This pride of possession may topple, however. When a dog is being scolded for peeing on the carpet, it looks for all the world as if it is ashamed. That would be a self-aware response.) We are aware of being aware. In other words, our level of self-consciousness transcends simple learning and memory in the brain.

Reductionist neuroscience does not explain how consciousness can allow us to separate ourselves from the activity of the brain. Reductionism gathers data and uncovers facts. In his research, Rudy wears a reductionist hat, since his primary field is Alzheimer's and the genes linked to that disease. But reductionist neuroscience doesn't explain who is actually experiencing the feelings and thoughts. There is a gulf between awareness and self-awareness. "I have been diagnosed with Alzheimer's" is a statement made from awareness; someone who is unaware would not notice that something is going wrong with their memory. "I hate and fear that I have Alzheimer's" comes from self-awareness. So the facts of the disease embrace all three states—unconsciousness, awareness, and self-awareness—without explaining how we relate to those three states. The brain is just doing what it's doing. It takes a mind to relate to that.

Of course, this "awareness of being aware" is also made possible by the brain. We do not claim to know, in reductionist terms, where awareness and self-awareness might be located in brain maps; they are likely not confined to one specific region. No one has solved this puzzle yet. While the brain produces feelings and thoughts that you identify with, super brain calls upon your ability to be the observer, or witness, who is detached from the thoughts and feelings delivered by the brain.

If a rage-aholic cannot stand back and observe what is happening when he has an outburst, then his anger is out of control. He is unconscious of where it comes from or what to do about it until a degree of detachment enters the picture. In brain scans, various centers in the cerebral cortex light up or grow dim, depending on whether someone has control over their emotions. But for many people, perhaps most, the thought of detachment from their emotions triggers a scary vision of a sterile, bland existence, devoid of passion.

But emotions change depending on how you are.

Unconsciousness: In this state, the emotions are in control. They rise spontaneously and run their own course. Hormones are triggered, leading too often to the stress response. If indulged in, unconscious emotions bring a state of imbalance in the brain. The higher decision-making centers are weakened. Impulses of fear and anger then have no controller. Destructive behavior may result; emotional habits become wired into fixed neural pathways.

Awareness: In this state, the person is able to say, "I am feeling X," which is the first step toward bringing X into balance. The higher brain offers judgment, putting the emotion into perspective. Memory tells the person how this emotion worked out in the past, whether for good or ill. A more integrated state follows, with higher and lower circuitry of the brain adding their input. When you stop being out of control emotionally and can say, "I am feeling X," you've reached the first step of detachment.

Self-awareness: When you are aware, you could be anybody. But when you are self-aware, you become unique. I am feeling X turns into What do I think about X? Where is it taking me? What does it mean? Someone who is angry can stop there, with almost no self-awareness. An irritable boss who chews out his subordinates year after year is certainly aware that he gets angry. But without self-awareness, he won't see what he is doing to himself and to others. He might come home one day and be flabbergasted that his wife has walked out. Once self-awareness dawns in you, the questions you can ask about yourself, about how you think and feel, have no limit. Self-aware questions are the keys that make consciousness expand, and when that happens, the possibilities are infinite.

Emotions are not the enemy of self-awareness. Every emotion plays its part in the whole; they are needed to attach meaning to events. Being emotional makes a memory stick in the mind. It's much easier to remember your first romantic kiss than to remember the price of unleaded gas that same night. Because they are "sticky" in this way, emotions are not detached. But detachment becomes part of the larger picture; it allows you to step back from your emotions (which is why every first kiss doesn't lead to a baby). This may sound coldly clinical, but detachment has its own joy. Once your experiences are not so sticky, you can transcend them to reach a higher level of experience where all of life is meaningful. By being mindful of your thoughts and feelings, you start creating new pathways that register not just anger, fear, happiness, and curiosity but all the spiritual feelings of bliss, compassion, and wonder. Reality making has no upper limit. When we assume that reality is a given, what we're really accepting isn't the world "out there" but our own limitations "in here."

How Ego Interferes

If self-awareness does have an enemy, it's the ego, which seriously constricts your awareness when it steps beyond its appointed function. That function is vital, as a glance at the brain immediately shows. While billions of neurons are remodeling trillions of synapses in an always evolving neural network, your ego leads you to believe that all is static and calm in your skull. This is not the case. Without a sense of constancy, you would be exposed to the brain's tumultuous process of reshaping itself as it responds to every experience you have, waking, sleeping, or dreaming. (The brain is highly active as you sleep, although much of this activity remains mysterious.)

Once new experiences have registered on the brain, your ego assimilates them. You are the *I* to whom new things are happening, adding to a storehouse of pleasure and pain, fear and desire that has been building up since infancy. Knowing that the brain's remodeling is always having an effect is important, even though your ego gives the illusion of constancy.

When Rudy and his wife Dora were raising their daughter Lyla, they decided that in her first year of infancy, Lyla would never be left to cry alone and unattended. Other parents criticized this decision, saying it would spoil the baby and turn Dora and Rudy into sleepless zombies, but they kept the promise they'd made to themselves. For Lyla, as for all of us, infancy lays the basic foundation of the neural network. Although the process takes place out of sight, a worldview is being shaped, and years later whenever a new experience of pleasure or pain occurs, it will be compared to the old ones before it finds its place in memory.

Dora and Rudy wanted to provide Lyla's brain with a basis of happiness, security, and acceptance, not discontent, abandonment, and rejection. Of course this approach required more work than attending the baby when it cried. But in infancy, a baby's whole world is her parents, and as she grew up, Lyla would have a deep-seated reason to view the world as accepting and nurturing. The world isn't fixed. It exists as we experience it and absorb it into our worldview. So the objection that Lyla would be unprepared for harsh reality wasn't valid. Like all of us, she will face the world according to the picture she has built up in her brain. (Lyla has turned out to be a very happy toddler who radiates the love she has been receiving.)

The ego is absolutely necessary for this function of integrating all kinds of experiences, but it is prone to go too far. *Egotism* is the common term for extreme self-centeredness, but that's not the issue here. Everyone is caught in a paradoxical situation with the ego. You can't function without one, but making everything personal can turn into ego delusion. "I, me, mine" overrides every other consideration. Instead of having a point of view and strong personal values (the good side of the ego), the egotist winds up defending his biases and prejudices just because he holds them (the bad side of the ego). The ego pretends to be the self. But the true self is awareness. When you shut out any aspect of experience by saying "That's not me" or "I don't want to think about it" or "This has nothing to do with me," you are excluding something from your awareness, building up an ego image rather than opening up to the endless possibilities of reality making.

Such narrow-mindedness comes at the price of reduced or imbalanced brain activity, which can be seen in brain imaging. New experiences equal new neural networks. They cause remodeling, which keeps the brain healthy. By contrast, when people tell themselves *I don't show my emotions* or *I don't like to think too much*, they shut down regions of the brain. The ego makes these rationalizations to constrict a person's awareness, which in turn constricts brain activity. Consider how some males equate *I am a man* with *A man doesn't show his emotions.* Leaving aside the rich existence that emotions provide, this attitude runs counter to evolution. The brain uses emotions to serve our instinctive needs, geared at ensuring survival. You must use your emotions to empower your passion for reaching personal goals. You must utilize your intellect to strategize, and finally you must detach your awareness to acquire the sobriety needed to achieve those goals. In other words, you need to cycle between the passion generated by your fears and desires and the rational thoughts associated with self-control and discipline. Charles Lindbergh had to possess drive and enthusiasm to attempt a record-breaking flight over the Atlantic, while at the same time being cool and objective enough to handle his plane while in flight. We are all like him.

The brain is fluid and dynamic. But it loses its balance when it is ordered to ignore or change its natural process. When you constrict your awareness, you constrict the brain and freeze your reality into fixed patterns.

EGO BLOCKS
TYPICAL THOUGHTS
THAT CONSTRICT YOUR AWARENESS

I'm not the kind who does X.
I want to be in my comfort zone.
This will make me look bad.
I just don't want to; I don't need a reason.
Let somebody else do it.
I know what I think. Don't try to change my mind.
I know better than you do.
I'm not good enough.
This is beneath me.
I'm going to live forever.

Notice that some of these thoughts make you look bigger while others make you look smaller. But in all of them, an image is being defended. The ego's true function is to help you build a strong, dynamic self (there will be more about how to do so in a later chapter), but when it intervenes to protect you unnecessarily, it is masking fear and insecurity. A middle-aged man who suddenly buys a red sports car might be feeling insecure, as could a middle-aged woman who pays for plastic surgery when the first crow's-feet appear around her eyes. But defending your ego is much subtler than that: the defenses we put up generally escape our notice. Instead of moving forward in the project of reality making, we wind up fortifying the same old reality that makes us feel safe. For some people self-importance is safe; for others it is humbleness. You can feel small inside and disguise it with an outward bravado, or you can take the same feeling and paper it over with timidity. There's no set formula. If you close off certain experiences, you don't know what you're missing.

But individual experience counts less than the brain's amazing agility in receiving, transmitting, and processing experiences. If you don't participate, the things you refuse to see will still affect you, but the effect will be unconscious. We've all known people who showed no grief when someone close to them died. Grief still had its way, but everything went on out of sight, an underground skirmish that continued despite the ego deciding "I don't want to feel."

Reality making is reciprocal. You make it, while it makes you. At the neurobiological level, excitatory neurotransmitters like glutamate are engaged in a constant yin-and-yang balancing act with inhibitory neurotransmitters like glycine, as your emotions and intellect play out the dance that creates your personality and ego. All of this provides you with a sense of who you are and what your response to life is at any given moment. Moreover, from your time in the womb, every sensory experience creates synapses, which consolidate your memories, laying down the foundation of your neural network. Those earliest-forming synapses were shaping you. Think of your response to a common household spider. Theoretically, you can have any response, but in reality your response is ingrained, and it seems natural once you have wired it in. *Spiders disgust me,* or *Spiders don't bother me,* or *I'm deathly afraid of spiders*—they are all personal selections that you have shaped but that also shape you. That's completely natural. The problem arises when the ego intervenes and turns a personal response into a fact: *Spiders are disgusting, spiders are harmless, spiders are frightening.* As statements of fact, these remarks are completely unreliable; they have turned a personal judgment into an "objective" reality.

Now instead of *spider,* insert the word *Catholics, Jews, Arabs, people of color, the police, the enemy,* and so on. Prejudice gets stated as a fact (*All those people are the same*), but at bottom there is fear, hatred, and defensiveness. Despite its subtle manipulations, the ego can be countered by a few simple questions. Ask yourself:

Why do I think this way?
What's really motivating me?
Am I just repeating the same old thing I always say/
   think/do?

The value of questioning yourself is that you keep moving on. You refresh your responses; you let self-awareness take in as much as possible. Having more to process stimulates the brain to renew itself, and the mind, with more responses at its disposal from the brain, expands beyond imaginary limitations. Whatever is fixed is limited; whatever is dynamic allows you to expand beyond limitation. Super brain is about removing limitations entirely. Each step brings you closer to your true self, which creates reality in a state of freedom.

SUPER BRAIN SOLUTIONS

OVERWEIGHT

Being overweight is a problem that is ripe for using the brain in a new way. More than one-third of Americans are overweight, and over a quarter are obese. Leaving aside medical issues, this epidemic is caused by the choices we make. In a society that consumes on average 150 pounds of sugar a year, eats one-tenth of all meals at McDonald's, and indulges in larger portion sizes every decade, you'd think our bad choices are so glaring that we'd rush to reverse them. But we don't, and massive public health advisories don't seem to help. Obesity has grown beyond reason because reason isn't effective in stopping it.

What is the baseline brain doing wrong? The culprit used to be considered moral. Being overweight was a sign of personal weakness, a holdover from the medieval inclusion of gluttony among the seven deadly sins. In the back of their minds, many overweight people accuse themselves of lack of willpower. If only they could stop being self-indulgent! If only they could stop punishing themselves with calories that fuel a vicious circle: eating puts on pounds, which leads to a worse self-image, and feeling bad about yourself gives you a good reason to console yourself by eating even more.

Decisions are conscious; habits aren't. In that simple statement, we begin to see being overweight from the brain's perspective. The unconscious parts of the brain have been trained to demand food that the higher brain doesn't want. The swing between overeating, remorse, and overeating has a counterpart in physiology. The hormones that serve as natural signals indicating that you have satisfied your hunger either get suppressed or are countered by other hormones that signal a ravenous appetite. Food itself isn't the issue. As tempting as a hot fudge sundae or a twenty-four-ounce porterhouse steak may be, they aren't addictive substances.

What is the issue, then? The answers, familiar by now, have the ring of futility. So many factors enter into diet and health that no matter where you turn, there's new blame to cast. People get fat, according to the experts, from

- Low self-esteem
- Poor body image
- A family history of obesity
- Genetic predisposition
- Early childhood training in bad eating habits
- Unhealthy fast foods and processed foods heavy in additives and preservatives
- The decline of whole foods
- Society's fixation on a "perfect" body that is unattainable by the vast majority
- The built-in self-defeat of constant dieting and yo-yo loss and gain

When you face such a discouraging pileup, the baseline brain gets quickly overwhelmed. This leads to a familiar pattern of self-defeating behavior. One failed diet leads to the next, out of sheer frustration and confusion. Failure breeds more frustration, but it also makes you prone to gimmicks and quick fixes: the unreasoning pressure of hunger, habit, and fantasies clouds your higher brain's ability to make decisions.

How can super brain change these entrenched patterns? First off, we need to declare a truce with fat. The baseline brain has not won the war. Studies show that many dieters lose weight, but up to 100 percent fail to keep the weight off for two years. Those who do keep a sizable amount of weight off relate that they are prepared to watch every calorie, day by day, for the rest of their lives. Brain chemistry plays its part. Dieters typically feel hungrier after they lose some pounds than before. Australian researchers believe the reason is a biological shift. The stomachs of successful dieters who then began to put the weight back on showed 20 percent higher levels of ghrelin, the so-called "hunger hormone," than before they went on the diet. A December 2011 report in the *New York Times* says, "Their still-plump bodies were acting as if they were starving and were working overtime to regain the pounds they lost." Your brain is responsible for regulating your body's metabolic set point through the hypothalamus, and dieting seems to affect that, too. People who come back down to a normal weight need 400 fewer calories a day than those who have consistently remained at their ideal weight over the years.

What the overweight person needs, in order to break out of self-defeat, isn't a new brain, a better metabolic set point, or balanced hormones. That is, the answer doesn't lie in these factors—they are secondary to something else: balance. An imbalance in the brain's circuitry results when the areas for impulsive behavior have been strengthened while the areas for rational decision-making have been weakened. The repetition of negative patterns also harms decision-making, because when you blame yourself or feel like a failure, lower parts of your brain are once again overriding your cerebral cortex. You restore mental balance by successfully making choices that are self-enhancing, such as when you stop reaching for food as an emotional fix. Once you restore balance, the brain will naturally tend to preserve it. This balance, known as homeostasis, is one of the most powerful mechanisms built into the involuntary or autonomic nervous system. The unique thing about the brain is that it runs on dual control. Processes run on automatic pilot, but if you instruct them to run the way you want, will and desire will take over. But this isn't a matter of willpower. Willpower implies force. You want to eat a second piece of pie or raid the fridge at midnight, but through sheer determination you resist.

That isn't will; it's resistance. Whatever you resist persists. There's the rub. As long as you engage in an inner war between what you crave and what you know is good for you, defeat is all but inevitable. In its natural state, will is the opposite of resistance. You go with the flow, and the will of nature, which has billions of years of evolution behind it, carries you. Homeostasis is the way your body wants to go; every cell has been engineered exquisitely to stay in balance (which is why, for example, a cell typically stores only enough food to last a few seconds. It has no need for extra storage because in the body's overall balance, every cell can count on being continuously nourished).

Super brain is about being in control of what your brain does. Our slogan is "Use your brain, don't let it use you." The area of weight includes patients suffering from eating disorders. A seriously anorexic girl can look in the mirror, see an emaciated frame with exposed ribs, elbows, and knees grotesquely sticking out, and a face that seems like a thin mask stretched over the skull. Nonetheless, what she sees is "I'm fat." The raw data entering her visual cortex at the back of her brain is irrelevant. As someone with an eating disorder, the body she sees is in her head. The same is true for everyone. The only difference is that we match a normal reflection in the mirror with a normal image in our heads. On the fringes of normal, millions of us see "too fat" when we look at a body that is comfortably within the range of normal. Of course, denial can set in, and after a certain point we may have too much extra weight to admit to. (A clever *New Yorker* cartoon has a woman asking her husband, "Tell me the truth. Does this body make me look fat?")

The key is to bring your brain into balance, then use its ability to balance everything—hormones, hunger, cravings, and habits. Your weight is all in your head because, ultimately, your body is in your head. That is, the brain lies at the source of all bodily functions, and your mind lies at the source of your brain.

Super brain requires you to relate to your brain in a new way. Most people are out of balance because their brains are so adaptable. The brain compensates for anything that happens in the body. Severely overweight people work around their obesity, leading normal lives within limits, raising families, enjoying loving relationships. At another level, however, they are miserable. Imbalance feeds more imbalance, perpetuating the vicious circle. They need to stop adapting to obesity and relate to the brain as the answer, not the problem.

WEIGHT LOSS AWARENESS

Stop fighting with yourself.

Ignore calorie counting.

Give up diet foods.

Restore balance where you know that your greatest imbalance exists (e.g., emotions, stress, sleep). Deal with the things that bring you out of balance.

Focus on reaching a turning point.

Let your brain take care of physical rebalance.

You can change a habit only in the moment when you feel the urge to act on your habit. Eating is no different. You find yourself reaching for pizza or sneaking ice cream at midnight. What's happening at that moment? If you can answer this question, you have an opening for change.

1. Either you are hungry or you are pacifying a feeling.

These are the two basic choices. At the moment you reach for food, ask yourself which one you are choosing.

I am hungry: If this is true, then eating is a natural bodily need, and it is fulfilled when hunger is no longer present (which is far short of being full or stuffed). A few hundred calories will fulfill a passing hunger pang. A meal amounts to roughly 600 calories.

I am pacifying a feeling: If this is true, then the feeling will be just as present as hunger. But you are in the habit of racing past the feeling. Or it may be disguised. Either way, stop and notice what you're feeling:

Overwhelmed and exhausted
Frustrated
Pressured
Distracted
Anxious
Bored
Insecure
Restless
Angry

Once you identify the feeling, name it to yourself, preferably out loud: for example, "I feel frustrated right now," "I feel exhausted right now."

2. Once you know what you're feeling, go ahead and eat.

Don't fight with yourself. The inner struggle between *I shouldn't be eating this* and *I have to eat this* never ends. If it had an ending, one side or the other would have won long ago. So register if you are hungry or are pacifying a feeling. Then eat.

3. Wait for an opening.

If you have been faithful about asking "What am I feeling?" before you eat, the time will come when your mind will say something new. *I don't need to eat this.* Or *I'm not really hungry, so why eat?* You don't have to anticipate such a moment. Certainly don't force it. But be prepared and be alert. Your urge to free yourself from a habit is real. It just happens, for the moment, to be not as strong as your eating habit.

When such an opening comes, act on your new urge, and then forget about it.

4. Learn better ways to cope.
When you pacify a feeling, it goes away temporarily, but it always returns. You are eating to cope with feelings. There are other ways to cope, and once you learn them, the urge to eat will lessen, because your body and mind will know that you aren't supplied with only one major coping mechanism.

Coping skills include:

Saying how you feel without fear of disapproval.

Confiding to the right person, someone who is empathic, nonjudgmental, and detached. (Confiding to people who depend on you for money, status, or advancement is never a good idea.)

Trusting someone enough to follow their guidance. Complete self-reliance is lonely and easily leads to distorted perceptions.

Finding a way to dispel the underlying energy of fear or anger. These two basic negative feelings fuel any addictive behavior.

Taking your inner life as seriously as your outer life.

Feeling good enough that you don't have to indulge yourself. Feeling bad is what tempts you to indulge yourself. It's not how good the food smells that leads you astray.

5. Make new neural networks.
Habits are mental grooves that depend on networks in the brain. Once set, they automatically respond. When a person fights the urge

to overeat, the brain is "remembering" that overeating is what it is supposed to do. It follows the groove automatically and powerfully. So you have to give your brain a new way to go, which means building new neural networks. You can't build them when the urge to eat hits you, but there are lots of times and lots of ways to build new brain patterns.

Nobody really enjoys having to pacify their feelings. It's too much like failure; it reminds you of weakness. But feelings don't want to be pacified, either. They want to be fulfilled. You fulfill your positive feelings (love, hope, optimism, appreciation, approval) by connecting with other people, expressing your best self. You fulfill your negative feelings by releasing them. Your whole system recognizes negative feelings as toxic. It's futile to bottle them up, divert them, ignore them, or try to rise above them. Either negativity is leaving or it's hanging on—it has no other alternative.

As you fulfill emotions, your brain will change and form new patterns, which is the whole goal.

You also need a reprieve from the inner struggle, conflict, and confusion that keeps your impulses, both good and bad, at war. This is where meditation helps. It shows your brain a place of rest. Leaving aside all spiritual implications, finding a place of real rest, where no aspect of your self is fighting with any other aspect, is immensely helpful. It gives your brain a foundation for change. In meditation you aren't following any grooves, patterns, or old conditioning. When your brain realizes this, it will want to experience it more. Therefore, instead of having old urges, you will begin to have more moments of balance, clarity, and freedom.

Your brain must become your ally. If it does not, it will remain your adversary.

Clarity is the key. What you see, you can change. What you can't see will continue to be with you. Since we never lose the ability to see, we are always open to change.

The goal of this program is not measured in pounds. In time, when you have trained your brain to recognize emotions, impulses, and the nonsatisfaction attached to overeating, you reach a turning point where you are confident about using your brain instead of letting it use you. You will easily choose not to overeat. With a clear purpose, you will naturally do what has always been good for you. We will meet these two themes—learning to use your brain instead of being its servant and learning not to force new behaviors—many times throughout this book. They are key principles for evolving toward super brain.

YOUR BRAIN
IS EVOLVING

All the beneficial choices you make are ways to evolve your brain. At a certain level, this is slow going; it took hundreds of millions of years for the most primitive animal brains to grow and develop into the spectacularly sophisticated human brain. In Darwinian terms, there is no other kind of evolution except this one, which depends on random mutations of genes over aeons of time. But we will argue that since people's choices create new neural pathways and synapses, along with new brain cells, human beings undergo a second kind of evolution that rests upon personal choice. Driven by what you want out of life, your personal growth works by reshaping your brain. If you choose to grow and develop, you are guiding your own evolution.

Super brain is the product of conscious evolution. Biology fuses with the mind. Up to the time you were around age twenty, nature took care of your physical development, which happened more or less automatically. You didn't choose to lose your baby teeth or learn to focus your eyes. But much else depended on the meeting of mind and genes. At three years old, most children are not ready to read. (An exceptional few are: a condition known as hyperlexia can bring on the ability to read before age two.) By age four or five, children are eager to read, and their brains are ready. A child discovers that black specks on a page mean something. Learning foreign languages also has its optimal age, which peaks in late adolescence.

Back when neuroscientists believed that the brain was fixed and stable, learning wasn't considered the same as evolution. But if the brain is changing as you learn, the two are synonymous. There was a news story recently about Timothy Doner, age sixteen, a high school student in New York City who decided to learn modern Hebrew in 2009, soon after his bar mitzvah. A tutor was hired, and the lessons went well. Timothy was discussing Israeli politics with his tutor, which led him to think about learning Arabic (considered one of the five hardest languages on earth), so he attended a college summer course on it.

The newspaper account continues: "It took him four days to learn the alphabet, he said, a week to read fluidly. Then he dived into Russian, Italian, Persian, Swahili, Indonesian, Hindi, Ojibwa, Pashto, Turkish, Hausa, Kurdish, Yiddish, Dutch, Croatian, and German, teaching himself mostly from grammar books and flash card applications on his iPhone." Timothy began posting videos online in other languages; he soon gained an international fan club. He discovered that he was a polyglot, someone who has mastered a number of foreign languages. Beyond this stage are hyperpolyglots, people obsessed with learning dozens of tongues. "Timothy was inspired by a video of Richard Simcott, a British hyperpolyglot, speaking 16 languages in succession."

That the prefix *hyper*, or "excessive," appears so often in this book (*hyperthymesia* for memory, *hyperlexia* for reading, *hyperpolyglots* for foreign language learners) attests to the low norms that we still set for the brain. But there is no reason to regard exceptional performance as excessive, a word that implies something freakish if not disordered. Our view is that we could be evolving into a new norm higher than ever before. Conscious evolution leads to super brain, which isn't freakish, disordered, or abnormal in any way.

Black specks on a page would have baffled our remote ancestors, but the brains of those early *Homo sapiens* were already evolved enough to permit language and reading. What they needed was time and the rise of cultures that would nourish language. What amazing things will we be doing routinely in the future, with essentially the same brain? Today we already lead lives of inconceivable complexity compared with people two generations ago.

Whose Face Is That?

The fact that Timothy could learn the fundamentals of a new language in a month, and even acquire a decent accent in Hindi or German, shows that the brain, when trained at the optimal time, can take a quantum leap with a skill that is already built into it. But what exactly has been built in? Science finds the answers one piece at a time, almost always as the result of a medical problem.

A striking example is face blindness, or prosopagnosia. Some soldiers returning home from World War II who had suffered head wounds failed to recognize their families' faces, or anyone else's. They could describe each feature precisely—hair color, eyes, the shape of the nose—but when asked at the end, "So, do you know who this is?" they answered with a baffled shake of the head.

At first, scientists connected face blindness to traumatic injuries; eighteenth- and nineteenth-century doctors had long noted queer mental deficits in their patients. But over the next five decades it emerged that face blindness can be a predisposition—just over 2 percent of the population seem to have it. In extreme cases, you cannot recognize even your own face. (The noted neurologist Oliver Sacks, who has written a book on the subject, has revealed that he has prosopagnosia. One time he apologized for bumping into someone, only to discover that he was apologizing to his own reflection in a mirror!)

Through injury or genetics, people with face blindness are defective in the fusiform gyrus, a part of the temporal lobe that is connected to recognizing not just faces but also body shape, colors, and words. Oddly, it can take years before someone discovers that they have this defect. Using an excuse like "I'm bad with faces," the person relies upon sensory cues, such as the sound of a friend's voice, or the way he dresses, in place of actually recognizing his face. One man reported that when his best friend at work changed her haircut, he walked right past her as if she were a stranger.

Prosopagnosia would seem to be a cut-and-dried diagnosis, traceable to a small, precisely located area of the brain. It's a well-documented fact that our brains are wired to allow us to recognize faces. Five visual patches in the back of the brain register sights unconsciously. For us to see them consciously, these signals must be relayed to the cerebral cortex in the front. When this circuitry doesn't work properly, recognition is absent. (Another specific patch allows you to recognize locations. When people have a defect there, they can describe a house in every detail but not recognize that they are standing in front of their own house.) Animals already possess the basic adaptation. Evolution has given them some incredible recognition abilities: Antarctic penguins returning home with food for their young can walk into a tightly packed flock of millions of birds and proceed directly to their own chick. (The standard explanation is that the parent has imprinted on the cry of its young, but other senses could be involved.) But there is another side to face blindness. Some people display the opposite ability—they are "super recognizers," an as-yet-little-studied phenomenon.

Super recognizers remember almost every face they've ever encountered. They can go up to someone on the street to say, "Remember me? You sold me a pair of black shoes at Macy's ten years ago." Naturally, the person being accosted almost never remembers. So startling are such encounters that super recognizers have been accused of stalking—being followed is an easier explanation to accept. Nor does the passage of time fool super recognizers. When shown pictures of seven- or eight-year-olds who grew up to be Hollywood stars, a super recognizer will instantly know whose faces they are. When asked how she does it, one woman shrugged. "To me, a face growing old is just changing superficially, like going from brunette to blonde or getting a new hair style." The deep wrinkles of an octogenarian don't mask the similarities to the same person photographed for a school picture in third grade.

If face blindness is a brain defect, what is super recognition? To answer that, we would have to know how people recognize faces in the first place. One thing we don't do is use cues, the way people with face blindness do to compensate for their disorder. When you meet a woman of a certain age, you don't go through a checklist of eyes, hair, nose, mouth, and then say, "Oh, it's my mother." You recognize her instantly—an ability that goes back to the predisposition a baby has almost from birth. If mothers are special cases, that doesn't make the mystery any easier. The brain forms complete pictures, known as gestalts, so biology underlies our ability to recognize faces all at once instead of one piece at a time.

The fact is that photons of light stimulating cells on the retina and signals being transmitted to the visual cortex carry no image. The optic nerve turns an image into a neural message that has no shape or luminosity. The information goes through at least five or six steps of processing. Light and dark regions are sorted out, outlines are detected, patterns decoded, and so on. Recognition comes very near the end of the process. But when you say, "Oh, it's my mother," nobody has the slightest idea how your brain has recognized her. The six stages of processing don't tell the story. Computer experts working in the field of artificial intelligence have tried to enable machines to recognize faces using various pattern cues. The results are rudimentary at best. If you see the photo of a familiar face that is slightly out of focus, you have no trouble knowing who it is, but even the smartest computer is stymied.

But if you take a photograph of a face and turn it upside down, you will lose the ability to recognize it, whether the face belongs to someone in your family, a celebrity, or even yourself. You can prove it to yourself by opening any celebrity magazine like *People* and turning it upside down—those famous faces will become indecipherable puzzles. But a computer built for facial recognition doesn't care if the image is upside down or right side up. It can easily be programmed for either one. Why did evolution give us the potential for super recognition but not upside-down faces?

Our answer wouldn't be brain specific. We'd say that the mind doesn't need upside-down recognition of faces, so the brain never developed it. A Darwinian would consider such a statement absurd. In strict Darwinian terms, there is no mind, no guidance of evolution, no purpose—nothing is inherited except through random mutations at the genetic level. It's quixotic for Rudy, as a genetic researcher, to allow mind into the equation. But he is convinced that the brain grows and develops in accord with what the mind wants. As evidence, we point to the fast-changing picture of the mind-brain connection. If neuroplasticity proves that behavior and lifestyle choices can change the brain, it's no great stretch to call this process evolutionary. As we evolve, variations slowly arise in our brains and genes.

At this stage of neuroscience, however, predisposition is a mixed picture with baffling aspects. We no longer consider nature to be separate from nurture in human development. In some cases nature dominates—some music prodigies begin tapping out Bach fugues on the piano at age two. But music can also be learned, which is nurture. The camp that wants all predispositions to be genetic has only part of the truth on its side; the opposite camp, which demotes inborn talent, claiming that ten thousand hours of practice can duplicate the ability of a genius, also owns only half the truth.

Let's go back to the polyglots who become fixated on learning dozens of tongues. To learn language, human beings depend upon genes, along with such vaguely defined traits as intelligence and attention; they also depend on nurture, which includes practice, necessary to train the brain in any new skill. But where do other necessary things fall, like patience, enthusiasm, passion, and even taking an interest? Must there be a gene for carving a cow out of butter for the Iowa state fair year after year? People develop very specific, even peculiar interests.

Far more mysterious is how a damaged or diseased brain can outperform a healthy one. That is the case with savant syndrome, now considered a form of autism but sometimes related to injury to the right temporal lobe. Those who suffer from savant syndrome (they used to be called "idiot savants") lack simple, everyday abilities but possess extraordinary ones. Musical savants, for example, can play on the piano any piece they've heard only once, including very complex classical music, even though they've never taken a piano lesson. Calendar savants can instantly tell you what day of the week any date falls on, including a date like January 23, 3323. There are language savants, too. One child suffering from this syndrome was unable to take care of himself or find his way unaided on city streets. On his own, he had somehow managed to teach himself foreign languages from books, which wasn't discovered until he got lost on a field trip. His caregivers panicked but eventually located the boy, who was calmly translating for two strangers, one of whom spoke Chinese, the other Finnish. Like Arabic, these are two of the five hardest languages on earth. Even more astonishing, the boy had learned Chinese with the textbook held upside down!

Spectacular examples like these can be daunting, but evolution is universal, open to all. The brain is unique among all bodily organs in being able to evolve personally, here and now. A five-year-old learning to read is evolving, as viewed from the physiology of the brain; he is laying down new pathways to give physical reality to the words of a Mother Goose rhyme. The adult brain is evolving when a person learns to manage anger, fly a jet, or develop compassion. The rich possibility of change demonstrates how evolution really works.

The Four-part Brain

Right now the scientific balance is tilted in the direction of brain over mind. Neuroscience uses the two words interchangeably, as if "I changed my mind" could be restated as "I changed my brain." But the brain has neither will nor intention; only the mind does. The brain also has no free will, even though the higher brain organizes choices and decisions. Neuroscience tries to simplify things by ascribing all human behavior to the brain. One sees journalistic articles about the "Brain in Love" and "God in the Neurons," which promotes the false assumption that the brain is responsible for love and faith.

To us, this is a mistake. When you hear static on the radio, you don't say, "There's something wrong with Beethoven." You know the difference between a mind (Beethoven's) and the receiver that brings that mind into the physical world (a radio). Neuroscientists are highly intellectual, sometimes brilliant people. Why don't they recognize such a basic difference?

A big part of the reason is materialism, the worldview that insists upon all causes being physical. Mind isn't physical, but if you sweep mind aside, you can study the brain on purely physical grounds. We hope we are making headway in convincing you that the brain exists to be used by the mind. Yet we must concede that evolution, working through genes, has structured the brain, giving you a receiving instrument broken down into definite parts. Our main thrust is that you can guide your own evolution, but on the way we must give credit to all the physical evolution that has already occurred.

For simplicity, we are going to divide the functions of your brain into four phases:

Instinctive
Emotional
Intellectual
Intuitive

These are the four ways that our minds work, as described by Satguru Sivaya Subramuniyaswami in *Merging with Siva*, a book that inspired and made a strong impression on Rudy as he was beginning to explore how ancient traditions of mind might relate to what we know about the brain today. For the human journey, evolution began with the instinctive parts of the brain (the reptilian brain, which is hundreds of millions of years old), then continued with the appearance of the part of the brain responsible for all emotions (the limbic system), and unfolded most recently to reach the higher functions of thinking (represented by the neocortex, which first appears in mammals and no previous animals). In humans, the neocortex forms 90 percent of the overall cortex. The neuroscientist Paul D. MacLean first proposed this "triune brain" in the 1960s. No one has successfully located the structure of the brain that supports intuition, and many neuroscientists would rather sweep the whole issue under the carpet. It is inconvenient for brain research that God is not, in fact, in the neurons; nor is art, music, a sense of beauty and truth, along with many other of our most valued experiences. However, since such experiences have been valued since the dawn of civilization, we include them in our four-part scheme. They must be, if we are to unriddle the brain at all levels of consciousness, from preprogrammed instinctive reactions to the visions of enlightened teachers who change the world.

The Instinctive Phase of the Brain

One-celled organisms that are billions of years old can respond to their environment; many, for example, swim toward the light. From these beginnings, the oldest phase of the brain evolved, the instinctive brain. It corresponds to behavior that is programmed by our genome expressly for the purpose of survival. Hundreds of millions of years of evolution have refined instinct. As massive as dinosaurs were, their behavior required only pea brains, no bigger than a walnut or apricot.

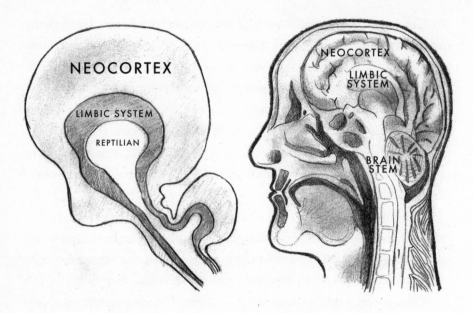

DIAGRAM 2: THE TRIUNE BRAIN

In the triune (three-part) model of the brain, the oldest part is the reptilian brain, or brain stem, designed for survival. It houses vital control centers for breathing, swallowing, and heartbeat, among other things. It also prompts hunger, sex, and the fight-or-flight response.

The limbic system was next to evolve. It houses the emotional brain and short-term memory. Emotions based on fear and desire evolved to serve the instinctive drives of the reptilian brain.

The most recent development is the neocortex, the region for intellect, decision making, and higher reasoning. As our reptilian and limbic brains drive us to do what we need for survival, the neocortex represents the intelligence to achieve our ends while also placing restraints on our emotions and instinctive impulses. Most important for super brain, the neocortex is the center for self-awareness, free will, and choice, making us the user and potentially the master of the brain.

Creatures that possess only this phase of the brain, like birds, can nonetheless display very complex behavior. Reptilian its brain may be, but an African gray parrot can mimic hundreds of words, and if current research is correct, it actually understands what the words mean. But if you gaze into the eyes of lizards and ostriches, frogs and eagles, you will detect no emotion. This vacancy can appear frightening, because we equate it with the merciless strike of a cobra or the pounce of predator on prey. Instinct preceded emotion on the evolutionary ladder.

The instinctive brain provides the natural impulses of the physical body that drive self-preservation, such as hunger, thirst, and sexuality. (When one writer referred to sexual craving as "skin hunger," his frankness was quite accurate in terms of the instinctual brain.) It includes entirely unconscious processes, too, like the regulation of the digestive system and the circulatory system—basically every bodily function that occurs automatically.

The anxiety that permeates modern society partially stems from our instinctive brain, tirelessly compelling us to pay attention to impulses of fear as if our survival depended on it. You won't die from a visit to the dentist, and other parts of the brain intervene so that fear doesn't impel you to jump out of the dentist's chair and run away. But the instinctive brain only knows how to pump out the impulse, not how to judge it.

If you observe yourself, you'll notice that the truce you've made with the instinctive brain is uneasy. Trying to ignore its impulses makes you restless, insecure, and anxious. Rudy remembers a time early in his college years, shortly after he lost his father to a heart attack. He was writing in his journal incessantly about the overwhelming feelings of anxiety and the cravings that dominate our teenage years. As postpubescent hormones surged, Rudy was baffled by his inability to ignore them. (The famous American food writer M.F.K. Fisher relays an anecdote about a man, grief-stricken by his wife's sudden death, who drove up and down the Pacific Coast Highway, stopping at every roadside diner and ordering a steak.)

Rudy intellectually knew that his anxiety-driven craving to go out and party with his friends all freshman year came from an irrational need for social acceptance, external validation, and stature among his peers. But he could not resist the urge to party when he should have been studying. Freshman year turned into a seemingly never-ending battle to somehow find the discipline to stay back at the library and study, while his instinctual brain won most of the victories.

Anxiety retained the upper hand until matters came to a head in 1979, during his senior year. It was New Year's Eve in Times Square. Rudy was part of the jostling crowd. The feeling in the air was palpably tense. The Ayatollah Khomeini in Iran was holding fifty-two Americans hostage. Bands of youths shouted curses against Iran and threw beer bottles. Rudy wandered away from his fraternity brothers and sat on the sidewalk, leaning against the rails of the subway entrance, feeling his anxiety peaking with the aggression all around him.

In such moments of personal crisis, just when the instinctive brain seems to have the upper hand, a radical shift may occur. Soldiers in battle may experience a sudden inner calm and silence as shells explode all around them. At this moment in Times Square, Rudy realized that all his anxiety was rooted in the basic impulses of fear and desire. Fear created doubt about how secure he was. Desire created appetites that demanded satisfaction, even when the circumstances were inappropriate.

Without yet knowing how the brain's circuitry is seamlessly integrated (the discovery lay decades ahead), Rudy could feel in himself that this was true. Fear and desire aren't strangers to each other—they are linked. Fear fuels the desire for activities that will alleviate fear; reciprocally, desire creates the fear that you can't, or shouldn't, get what your appetites demand. We turn to scientists and poets to validate the conflicts that the instinctive phase of our brain creates. Freud spoke of the power of unconscious drives for sex and aggression; these nameless forces are so primitive that he labeled them *id* ("it" in Latin). Id is powerful, and Freud's slogan for curing his patients was "Where Id is, Ego shall be." The world constantly witnesses the destructive power of our primal drives. Fear and aggression are waiting to storm the gates of reason.

Shakespeare looked at himself running after women and called lust "the expense of spirit in a waste of shame." That sonnet could serve as a lesson in brain anatomy, since it maps out the conflict between impulse and reason.

> *The expense of spirit in a waste of shame*
> *Is lust in action; and till action, lust*
> *Is perjured, murderous, bloody, full of blame,*
> *Savage, extreme, rude, cruel, not to trust.*

There could hardly be a more accurate description of primitive drives and how people behave when sex overwhelms everything else. If two bighorn sheep butting heads in rutting season wrote poetry, they'd describe their ungoverned urges like this. But being human, Shakespeare looked back upon lust with remorse:

> *Enjoyed no sooner but despised straight;*
> *Past reason hunted; and no sooner had,*
> *Past reason hated, as a swallowed bait.*

He compares himself to an animal that has been lured by bait set in a trap. The satisfaction of lust has brought a new perspective, one of self-reproach. (We have no evidence that Shakespeare had a mistress, but he was a married man who had fathered a daughter and newborn twins when he left his family behind in Stratford to seek his fortune in London in 1585.)

Why was the trap laid? Shakespeare doesn't blame women. He says the trap was laid by our nature, to drive us mad:

Mad in pursuit, and in possession so . . .
A bliss in proof, and proved, a very woe.

He has stepped from the instinctive brain into the emotional brain, which evolved next. Elizabethan poets were always in some high passion, whether of love or hate. But Shakespeare has indulged his feelings enough, and now the higher brain is invoked. It looks at all this mad behavior and delivers a sad moral:

All this the world well knows; yet none knows well
To shun the heaven that leads men to this hell.

In moments when we are divided, the brain can physically represent every aspect of our mental war. To Rudy, at that moment in Times Square, the setup that causes fear and desire to rule behavior seemed crystal clear. The unruly street kids yelling at Iran and smashing bottles were also himself, even though he was a passive bystander. Fear and desire drove them. An instinctive desire for power and status, as any good psychologist will tell you, creates anxiety driven by the fear of rejection and loss of power. Too intense a desire for success leads to stronger fears of failure, and if fear rises, it can create failure. The instinctive brain traps us between wanting something too much and not getting it at all.

As with any phase of the brain, instincts can go out of balance.

If you are too impulsive, your anger, fear, and desire will run out of control. This leads to rash actions and regret afterward.

If you control your impulses too much, your life becomes cold and repressed. This leads to a lack of bonding with others and with your own basic drives.

ESSENTIAL POINTS: YOUR INSTINCTIVE BRAIN

See that instincts are a necessary part of your life.

Be patient with fear and anger, but don't indulge them.

Don't try to argue yourself out of your impulses and drives.

Don't repress thoughts and feelings out of guilt.

Be aware of fear and desire. Awareness helps to balance them.

Just because you feel impulsive, don't always act on impulse. Higher parts of the brain must be consulted, too.

SUPER BRAIN SOLUTIONS

ANXIETY

Anxiety creates a false picture of the world, piling on things to be afraid of that are in fact harmless. The mind adds fear. If the mind can undo the perception of fear, the danger will vanish.

To begin with, life cannot exist without fear, and yet fear creates paralysis and misery. The two aspects, one positive, the other negative, meet inside your brain. For people who suffer from free-floating anxiety (one of the most common complaints in modern society), the short-term solution is a chemical fix-it: tranquilizers. We've already warned about the flaws of chemical fixes, in terms of side effects, but the most basic problem of all is that drugs don't cure mood disorders, including anxiety. Just as being sad is universal while depression is abnormal and unhealthy, fear is universal while free-floating anxiety gnaws away at the soul. As Freud pointed out, nothing is more unwelcome than anxiety. Medical studies have found only a few things that the mind-body system cannot adapt to: one of them is chronic pain, the kind that gives no remission (shingles, advanced bone cancer), and another is anxiety.

Free-floating means the thing you fear is not a specific threat. In the natural scheme, our fear response is physical and targeted. Victims of crime report that during the act, as the weapon of their assailant loomed large in their visual field, they went into a state of hyperalertness, their hearts racing. These aspects of the fear response come automatically from the lower brain, and the things that cause you worry and anxiety are thought to be programmed in the amygdala. That doesn't tell us enough, however. Once you become anxious in a pervasive sense—as happens, for example, to chronic worriers—the whole brain gets involved. Fear is targeted and specific; anxiety is pervasive and mysterious. People who suffer from it don't know why.

What they experience is like a bad smell that stays on the edge of their awareness no matter how hard they try to pretend it isn't there. To heal the anxiety, they can't attack it as one thing; the bad smell has seeped everywhere. In other words, their reality making has gone awry. Anything or nothing can trigger anxiety in them. They always have something to be afraid of, a new worry or threat. To find the solution, they must learn not to fight the fear but to stop identifying with their fears.

Achieving detachment is only possible if you can get at what makes fear so sticky. In its positive, natural state, fear dissipates after you run away from the saber-toothed tiger or kill the woolly mammoth. There is no psychological component. In its negative, pervasive state, fear lingers. Its stickiness has a number of aspects.

HOW ANXIETY BECOMES STICKY

The same worry keeps returning. Repetition makes the fear response stick in the brain.

The fear is convincing. When you believe in the voice of fear, it takes over.

The fear stirs a memory. What you fear resembles something bad in your past, which brings back the old response.

Fear leads to silence. From shame or guilt you don't speak your fear, so it festers.

Fear feels bad, and you shove the pain out of sight. But repressed feelings endure. What you resist, persists.

Fear is crippling. You feel too weak to do anything about it.

Earlier we talked about the depressed response in terms of a behavior that turns into a habit. That's one way to describe stickiness when it comes to emotions; the points we made about how depression turns into a habit are worth going back to, since they also apply to anxiety. What we are adding here is the multidimensional aspect. Fear reaches out with many tentacles, and every attachment is unhealthy. To undo fear, we need to break its reality down. Each part, taken on its own, is manageable. You can dismantle it for the simple reason that you are at the center of reality making.

1. The same worry keeps returning. Repetition makes the fear response stick in the brain.

Repetition deepens the rut that keeps any response fixed. If you have to walk through a dangerous part of town at night after work, doing it over and over makes the threat feel worse. Sometimes you grow used to it. Children who live with angry parents can predict fairly well when to expect another blowup. But repetition is never simple. The same children will find, usually many years later, that the abuse that their angry parents dole out affected them badly. In the case of anxiety, they internalized the repetition. You turn into the abuser, delivering the same "be afraid" message over and over.

It helps to realize that you are playing this double role of abuser and abused. Chronic worriers can't see it. They repeat the same worries (*What if I didn't lock the house, what if I lose my job, what if my child is on drugs?*) and actually think they are being helpful. The irritated reactions of family and friends don't end the delusion. If anything, the worrier steps up her worry because no one else is paying attention. It's her responsibility, then, to worry for everyone else.

The mind, trapped inside itself, can't look far enough to see that chronic worry does no good. It doesn't recognize the repeated, obsessive assault of fear as negative. It becomes a kind of fix. You endure a little nagging pain in order to ward off huge threats that could bring disaster. A kind of magical thinking is involved, also.

The worrier is chanting a kind of incantation that is supposed to keep the threat away. (*If I worry about losing all my money, maybe it won't happen.*)

To end the influence of repetition, awareness must come into play, by consciously thinking thoughts like the following:

I'm doing it again.
I feel bad when I worry.
I need to stop at this moment.
The future is unknown. Worrying about it is pointless.
I'm doing myself no good.

A woman caught up in a bad marriage was afraid for herself, constantly worrying about her future. She feared being alone. She was afraid that her children would side with her husband, that he would destroy her good name with their friends, and that her work would be affected. A state of high anxiety resulted. She assaulted herself every day with a mounting crescendo of worries.

But the facts spoke otherwise. Her children and her co-workers loved her. She did fantastic work. Her husband, although he wanted out of the marriage, provided a large settlement without complaint. He wasn't even bad-mouthing her or forcing their friends to take sides. The actual problem was far simpler than it looked: She became anxious every time she thought about the future. Luckily, she had a confidante who had insight into this pattern. No matter what worry the anxious woman brought up, her confidante said, "You get afraid every time you think about the future. Just stop. I've known you a long time. The things you worried about two years ago, five years ago, ten years ago, all worked out. They will this time, too."

Of course, this reassurance didn't sink in at first. The woman's repetitive worry had become a habit; by bringing the same warnings to mind over and over, she felt she had a kind of control over fear. But her confidante persisted. No matter how anxious the woman acted, her confidante would say, "You get afraid when you anticipate the future. Stop it." Several months passed, but eventually the tactic worked.

People who are stuck in self-destructive worrying know the old pattern doesn't work from the get-go. They break out of it not by learning to stop the mental process but by overriding it with an emerging awareness that says, "The fear isn't real. I'm the one creating it." The anxious woman became aware that she was abusing herself through self-induced fear. She learned to stop herself when the merry-go-round of worry started to whirl.

2. The fear is convincing. When you believe in the voice of fear, it takes over.

If you think that something is true, it sticks with you. This goes almost without saying. We all want to believe the words *I love you* if they come from the right person; the memory can reassure you for years, if not a lifetime. But being convincing isn't the same as being true. Suspicion is a prime example. If you suspect that your spouse is cheating on you, no amount of proof that he or she is not will persuade you otherwise. You are too convinced by your suspicions. Jealousy is suspicion taken to a pathological extreme. When lovers are in its grip, all are unfaithful, and when such stickiness exists, the actual facts might as well not exist.

Anxiety is the most convincing emotion of all, in part because evolution has hardwired the brain to react with the fight-or-flight response. If you are in battle staring into the mouth of a cannon, your racing heart tells you in no uncertain terms what you must do. But when the condition is free-floating anxiety, the voice of fear isn't telling you the truth. It is using the power to convince you, even when you have nothing to be afraid of. Detachment has healing abilities here. If you can say to your fear, "I don't believe you. I don't accept you," its power to convince will diminish.

Here the mind must lead the brain. If the brain is exposed to a terrible outside event (e.g., an airplane crash, a terrorist attack), it reacts with fear, but pictures of that event or any other strong stimulus that brings it to mind will evoke the same reaction. Reflex reactions speak to us; they have a voice. But the mind exists to sort out the real from the unreal. When the mind is leading the brain out of anxiety, it has thoughts like the following:

> Nothing bad is happening to me. I can handle the situation.
> Worst-case scenarios are extremely unlikely to occur. This isn't one.
> I am not alone. I can turn for help if I need it.
> My anxiety is just a feeling.
> Does this feeling make sense?
> Things are okay, and I am okay, right now.

By putting the voice of fear in its place this way, you make it less convincing. Each time you do it, repetition comes to your aid instead of undermining you. Every realistic appraisal makes the next one easier. Anxiety has no power to convince when you see that reality doesn't match your state of alarm.

3. The fear stirs a memory. What you fear resembles something bad in your past, which brings back the old response.

Reality making occurs here and now, but no one lives in isolation. As much as you try to live in the present, your brain stores and learns from every experience by comparing it with your past. Memory is immensely helpful—it enables you to get on a bicycle and ride without having to learn how every time. This is the natural, positive use of memory. The destructive side, which fuels anxiety, makes you a prisoner of the past. Impressions of old wounds and traumas shouldn't have such a strong psychological component. But they do; hence their stickiness. (As Mark Twain wittily put it, "The cat, having sat upon a hot stove lid, will not sit upon a hot stove lid again. But he won't sit upon a cold stove lid, either.")

For *cat*, substitute the word *brain*, because it is just as trainable. Once it is exposed to a painful experience, the brain gives a privileged pathway to remembering the pain if it should come up in the future. It's a useful evolutionary trait, which is why a small child doesn't put its hand in the fire more than once. But the reflex is thoughtless, so old memories get blurred into present experience where they don't belong. For example, child psychologists make a distinction between telling a child what to do and telling him what he is. The child easily forgets the first kind of statement—which of us remembers to look both ways before crossing the street? But the second kind of statement lingers. Once a child is told "You're lazy" or "Nobody will ever love you" or "You're just plain bad," he will grow up with those words in his head, often for life. We rely on our parents to tell us who we are as small children, and if what they say is destructive, there is no escape without consciously healing the old memories.

Bringing awareness to the stickiness of memory requires new thoughts like the following:

I'm acting like a child.
This feeling is how I felt a long time ago.
What could I feel now that fits the situation better?
I can view my memories like a movie without buying into
   the story they tell.
All that I'm scared of is a memory.
What's actually in front of me?

Memory is the ongoing story of your life, and it does no good to keep reinforcing the story unconsciously. You need to step in and add something new, however small. Memory is incredibly complex, but it tends to trigger a simple reaction:

A is happening.

I remember B, something unpleasant in the past.

I'm having reaction C, just the way I always do.

This simple pattern recurs in all kinds of situations, such as going home for Christmas, seeing a politician on TV from the opposing party, or getting caught in a traffic jam. Know that even if you have no control over event A and memory B, reaction C will open up the chance to intervene. While having your reaction, you can work on it, examining your response, moving the negative feelings that are evoked, and not running away until you feel that you have had the response you want to have. In the chain reaction, A, B, and C can hit all at once, but even so, you can intervene consciously to break the chain, and when you do, memory won't be quite so sticky anymore.

4. Fear leads to silence. From shame or guilt you don't speak your fear, so it festers.

There's an old-fashioned nobility about keeping your fears to yourself. Males in particular are reluctant to admit that they're afraid, for fear of not being masculine enough in the eyes of other men. Women are more likely, thanks to social acceptance among other women, to speak about their emotions. But sharing, too, has its pitfalls, since people are pressured to keep their confessions or complaints within socially accepted boundaries. The most difficult things, colored by guilt and shame, rarely get expressed.

We shouldn't be surprised, therefore, that more often than not, abused children keep quiet and suffer in silence. Child abuse relies upon this reluctance to speak out. The victim feels that she must have done something wrong simply by the fact of being victimized. Switch to anxiety as the problem rather than abuse, and you see that the mind plays a double role: it accuses a child of doing something wrong and at the same time it tells her that she is being violated, which puts the blame on the abuser. This is a double bind. Let's look closer at how such a trap works to paralyze the child. Suppose a mother is angry at her child and wants to spank it, and she says, "Come to Mommy," with a coaxing smile. The child hears the words but at the same time sees that Mommy is angry and is going to mete out punishment. Two contradictory messages clash, which is a double bind.

Speaking out your fear untangles the bind. A young child who doesn't want to be spanked may simply balk and refuse to move. He isn't old enough to say, "You're making me feel afraid even though you're pretending to be nice." If you feel anxious, it's up to you to untangle your feelings yourself, but by definition speaking out your fear requires another person. You need more than a listener. You need a confidante, someone who has been through the same kind of fear. Such a person must be at least a few steps ahead of you. They need to empathize and show you that fear can come to an end. In other words, they have walked the walk when it comes to anxiety. Well-meaning friends are not necessarily good at this. They may respond by judging against you, taking the side of guilt and shame. ("You wished your baby was never born? Oh my God, how could you?")

Emotional maturity begins with knowing that thoughts aren't actions. Having a bad thought isn't the same as carrying it out. Guilt doesn't recognize the difference. Therefore, to come out of silence, you have to learn, by watching another person's reaction, that it's all right to have any thought you want. The point is to get out of the anxiety that the thought induces. To get to the point where you can find such a mature confidante, you need to cultivate thoughts like the following:

I don't want to live with my guilt.
Silence is making it worse.
No matter how long I wait, my anxiety isn't going away on
    its own.

There is someone who has been where I am.
Not everyone will feel as bad about me as I do. There even
   might be someone who wants to sympathize.
The truth has the power to set me free.

One of the more peculiar findings in psychiatry is that people who are on the waiting list to go into therapy often improve before they have the first session, and the improvement can be as much as they can expect to receive from a psychiatrist. Before working up the courage to go into therapy, these troubled people overcame the pressure from inside to keep silent. That step, in and of itself, has the power to heal.

5. Fear feels bad, and you shove the pain out of sight. But repressed feelings endure. What you resist, persists.

Avoiding pain is effective. Humans aren't lemmings. If your friends dare you to jump into an empty marble quarry, you don't have to just because they do. But the simple tactic of pain avoidance backfires in the brain. You've probably heard the old challenge, "Try not to think of an elephant." The mere mention of "elephant" triggers associations in the brain. This is essential for human existence—it's how we learn, by steps of association. At this moment you associate the words on this page with all the words you've ever read, and thus you can decide to absorb and accept what you're reading or not.

Fear associates pain with pain, however. Its associations feel bad, and when somebody mentions it, you will try very hard to push pain out of the way. Freud, among many other students of the mind, believed that pushing feelings, memories, and experiences out of sight, which is called repression, doesn't work. Lurking somewhere out of sight are the associations you don't want to face. Carl Jung, following Freud's lead, believed that part of us creates a fog of illusion in order to keep life from being too painful. He called "the shadow" all the hidden fear, rage, jealousy, and violence that gets shoved into secret compartments in the psyche.

On the face of it, Freud seems to be wrong; most people are quite good at denial. They don't face painful truths. They block out all kinds of experiences they wish they'd never had. But the shadow taps out messages in the dark. Repressed feelings rise up like ghosts. Sometimes you feel anxious because your fear is trying to rise up. But repression is tricky. You can feel anxious because you are worried about keeping secrets; or because you know that one day you will be exposed; or because the pain of avoiding pain is too great.

The antidotes to repression are two: openness and honesty. If you are open to all your feelings and not just the nice ones, you don't have to repress anything. You have no dirty little secrets to squirrel away. If you are honest, you can name your feelings, no matter how unwelcome they are. But nobody is perfect at this. Freud announced to a shocked world that all infants are hiding a sexual attraction for their mother or father. If that is a universal secret (it very well may not be), then repression is epidemic. We don't have to settle that deep psychological issue here. The important thing is to heal. In order to find the courage to bring up your secrets, you need detachment. A one-year-old who wets the bed is detached because no guilt is attached to wetting the bed at that age. A four-year-old who does the same thing and gets scolded for it will try to hide the next incident. A forty-year-old who wets the bed can enter into very convoluted states of embarrassment.

To speak the feelings that you have suppressed for years, your biggest risk is that the person in whom you confide will react judgmentally, at which point you may wish you had kept your secret hidden. But then, guilt has a nasty way of making us turn to the wrong people when we want to bare our souls. That's because we still play the double role of abused and abuser. We don't seek someone who then turns out to judge us: we seek them out *because* we know they will judge us. So you must prepare the ground first, with thoughts like the following:

I know I'm hiding something, and it hurts.
It's scary to come clean, but that's how I will heal.
I want to be unburdened.
Being haunted makes me too anxious.

When you are keeping secrets, especially secret emotions that you judge against, it's hard to realize that forgiveness is possible. The state of forgiveness is too far away; it feels imaginary compared to the anxiety you feel here and now. Just remember that forgiveness is the last step, not the first. You approach it step by step. Your responsibility to yourself is only to want to forgive yourself and then to figure out the next step, however small, to get toward healing. The first step could be reading a book, keeping a journal, or joining a support group online. Whatever it is, the point of taking a first step is always the same. You stop heeding fear and learn to accept your feelings for what they are: natural events that belong in your life.

6. Fear is crippling. You feel too weak to do anything about it.

When someone is frightened, they can become paralyzed with fear. Two soldiers charging up the hill at Gettysburg or two firefighters facing a burning house may feel the same fear, as measured physically by changes in their brain. But if one is a veteran soldier or firefighter, their fear doesn't immobilize them. They relate to it in a different way from the soldier who has never faced gunfire or the rookie firefighter who has never run into a blaze. Being frozen with fear, in other words, depends on more than the body's fear response.

Fear's ability to freeze you in your tracks is mysterious and changeable. An experienced rock climber can be enjoying a normal day's climb, with nothing especially risky ahead of him, when suddenly he can't move another inch. He freezes on the rock face, because suddenly his mind, instead of taking for granted the danger of falling, thinks, *Oh my God, look at where I am.* The raw fear of falling takes hold, and it doesn't matter how often the climber has been on the same rock face. He has computed the experience in a new way.

How you choose to reinterpret any bit of raw input can work to your advantage. That's how you decided to stand up to a bully on the playground or got back on the horse after it threw you. Since your brain isn't you, neither is its reactions. FDR was making a universal statement when he declared that "the only thing we have to fear is fear itself." The way out of any fear is to get past its power to frighten you. (Because economists don't factor fear into their equations, many were baffled at the sudden and total collapse of the American economy after the housing bubble burst in late 2008 and banks started to topple. According to the data at hand, the economy was strong enough not to lose as many millions of jobs as it did. But this was a case where data didn't matter. People allowed themselves to be frightened by fear. Manageable anxiety was transformed into panicky behavior.)

Mind, brain, and body are seamlessly connected. Being afraid of fear leads to all kinds of symptoms, such as muscle weakness, fatigue, loss of enthusiasm and drive, forgetfulness that you were once unafraid, poor appetite and sleep—the list goes on. Imagine that you find yourself hanging from a cliff by your fingertips, and it's midnight. In the pitch darkness, you are terrified of falling hundreds of feet to your death. Then someone leans over and says, "Don't worry, the drop is only two feet." Suddenly you relate to your fear response in a new way. It's easy to feel the panic and helplessness of hanging from a cliff, but when fear lifts, the whole body changes. Even if your fear lingers, knowing that you are safe signals the brain to restore you to your normal state.

Anxiety tells you that you are in great danger, and the body doesn't operate with a rheostat to turn the fear response up and down—it only knows on and off. Even fear of the number thirteen, technically known as triskaidekaphobia, can feel as if you are going to die. A blunt but effective treatment for phobias uses saturation to short-circuit the exaggerated fear.

A patient was deathly afraid of rat poison and electrical cords. The sight of either threw him into a panic; during these attacks his fear made him mindless. His therapist put him in a chair and sedated him. When he nodded off, he was draped with empty boxes of rat poison around his neck and wrapped in electrical cords. As soon as he woke up and saw what had happened, the patient screamed bloody murder. As far as his phobic reaction was concerned, he was about to die. Phobics will do anything to avoid this feeling, but here he couldn't escape it. He went into a frenzy of fear. But as the minutes passed and he didn't die, he found an opening. The phobia was no longer in total control because he wasn't in total terror of it.

We are not recommending saturation; that's not our message. But it is necessary to defuse the fear that fear provokes.

In order to get over your fear of being anxious, you need to cultivate thoughts like the following:

I am not going to die, no matter how scary this is.
I need to face my exaggerated sense of danger.
Since I know I can survive, I can risk not running away
    from my fear.
I can face fear and still do things that scare me.
The more I face fear, the more control I gain over it.
When I am really in control once more, my fear will vanish.

This is the final step in dismantling the stickiness of anxiety. You can approach the problem, however, by starting with any of the steps we've covered. The aim is always the same: to get to a more detached place. Phobias prove that reality isn't strong enough to conquer fear.

You can put a few harmless spiders on someone who is deathly afraid of them, and their panic may induce a heart attack. What's stronger than reality? Knowing that you are the reality maker. That is the pivotal point. Once you regain the clarity that comes from knowing how reality is made, you're free. You've invaded the workshop of the brain and declared that you are in control. The creator has returned.

THE EMOTIONAL BRAIN

Fear and desire are bred in your instinctive brain, mediated by your emotional brain, and negotiated by your intellectual brain. These structures meet the mind's demand to process lust, infatuation, anger, greed, jealousy, hatred, and disgust. All such feelings are tied to survival in the course of evolution. The fight-or-flight response in reptiles implies a brain with fixed circuits for that response. Humans didn't evolve to be rid of the same circuits, or even to nullify them (the way, for example, the tails of early mammals shrank to a vestigial bone at the end of our spine).

Instead, the human brain has added layers of new upon old. (In the case of the cerebral cortex, the outermost layer of the brain, the layers are quite literally like bark on a tree. *Cortex* means "bark" or "rind" in Latin.) This layering keeps integrating what came before rather than throwing it away. While past memories of pain and discomfort drive fear, memories of past pleasures and enjoyment drive desire. Evolution pushes and pulls at the same time. It is impossible to say where wanting pleasure ends and avoiding pain begins. Shakespeare might have been ashamed of his lust, but he didn't ask for it to be taken away. The emotions based on fear and desire work hand in hand with each other. For example, your fear of rejection by your social group dovetails with your desire for power and sex, sustaining the individual and the species at the same time.

Emotions feel as urgent as instincts, but a new development is taking place. Freud called instinctual drives *it* because they were too primitive to name. Emotions have names, like *envy, jealousy,* and *pride.* When a poet declares that love is like a red, red rose, he's expressing our fascination with naming our emotions and building an entire world around them. So emotions are a step in the direction of awareness.

The conflict between instincts and emotions teaches us that humans have evolved—with much pain and confusion—to learn. You must be mindful of your fears and desires. They have no control built into them, and neither does the reptilian brain. The complicated limbic system is our center for emotion, but also for obscurely related things like long-term memory and the sense of smell. Smelling a perfume or chocolate cookies is enough to bring memories flooding back from the past (in the case of Marcel Proust, it was dipping a madeleine cookie into tea) because the limbic system unites smell, memory, and emotion. It evolved second, after the reptilian brain, but still early. All four-footed animals, including the earliest amphibians, seem to have a developed limbic system. Emotion, unlike smell, may be a recent development in the story. Or perhaps emotions couldn't exist until language gave them names.

Our tendency to look down at the lower brain for being primitive is a mistake. You can "smell" trouble, with the kind of certainty that the higher brain envies. The lower brain has no doubts or second thoughts. It can't talk itself out of what it knows. No one speaks of the wisdom of the sex drive, but our instinctively driven emotions are definitely wise. They stand for the kind of awareness that leads us to be happy. Before the word *geek* was invented, universities started to attract the kind of obsessive young males who were brilliant at writing computer programs. They sat up night and day writing code. The digital age was built on their midnight oil. But there tended to be a fast turnover in these twenty-somethings, and when asked why, the dean of a leading university sighed. "We can't keep them from crossing the quad, and as soon as they run into a girl, they vanish."

The loss to binary codes is humanity's gain. With the emergence of the emotional brain, awareness began to pry itself loose from physical survival. The various areas of the limbic system, such as the hippocampus and the amygdala, have been precisely mapped, and they can be correlated to all kinds of functions through fMRIs. If this precision tempts neuroscientists to claim that the limbic system is using us for its own purposes, the way instinct does, the claim needs to be resisted. The instinctive brain, because it evolved for survival, needs to use us. Who wants to choose to digest his food after each meal? Who wants to see the car ahead swerve out of control and have to think for a moment before reacting? Huge areas of life should be on automatic pilot, and therefore they are.

But emotions, even as they well up spontaneously, mean something, and meaning is a department we all want to be in charge of. "I can't help it. Every time I see the ending of *Casablanca*, I cry," someone may say. Yes, but we choose to go to the movies, and one reason is to feel strong emotions without risk. It's okay for a man to cry at the ending of *Casablanca* or when Old Yeller is shot, even one who believes that grown men don't cry. Movies are vacationland for the limbic system—not because the brain needs to cry, but because under the right circumstances, we need to cry. The emotional brain feels no emotions. You feel emotions while using it.

Wrapped up in the emotional phase of the brain, however, is a new conflict, one we've already touched upon: memory. Memory is the most powerful way to make emotions stick, and once stuck, they are difficult to remove. We've already discussed the stickiness of one emotion, anxiety. In Sanskrit, the stickiness of experience is called *samskara*. It is defined as the impression left by past actions, or karma. These are exotic words, but every Eastern spiritual tradition is rooted in a universal dilemma: the struggle to break the grip of old conditioning, which creates pain today by remembering the pain of yesterday. The process of laying down karmic impressions is an inextricable aspect of the emotional brain.

Whether you believe in karma or not makes no difference. You are laying down impressions in your nervous system all the time. Every like and dislike you have (*I hate broccoli, I love asparagus. I hate her, I love you*) is due to past impressions. This is more than data processing. Anyone who compares the human brain to a computer should be asked if computers like broccoli or hate fascism. Emotions guide preferences, and computers are devoid of emotion.

Since laying down impressions comes effortlessly, you'd suppose it would be easy to remove them. Sometimes it is. If you misspeak, you can correct yourself with "Forget what I just said," and your listener will. But impressions that make a lasting difference cannot be removed even with the greatest effort. Trauma stays with you. Because memory is so poorly understood, its footprints cannot be detected in the limbic system. Yet somehow vivid memories are sticky by nature.

You need to have an open emotional life and to value your feelings. But when emotions gain the upper hand, there is more evolving to do. In particular, we believe that you should be a witness to your emotions. This doesn't mean that you should simply stand by and watch yourself get mad or go into a panic, should those emotions spring up. Emotions want to run their course; like instincts, they want what they want. But you shouldn't fuel them to excess. Anger, for example, is already hot and raging. It doesn't need you to throw kerosene on it. By observing your anger, you create a small gap between you and your emotion. If you observe, *This is me getting angry*, the *me* and the anger are now separated. In that tiny act of detachment, the emotion loses momentum. You always have the choice to use any part of your brain as your partner. The terms of the partnership are up to you.

As with any phase of the brain, emotions can go out of balance.

If you are too emotional, you lose perspective. Your feelings convince you that they are the only things that matter. Excessive emotion is exhausting and depletes the whole mind-body system. If you indulge your emotions long enough, you become their prisoner.

If you control your emotions too much, however, you lose touch with how your life feels. This leads to the illusion that intellect alone is enough. Ignorant of how powerful a hidden emotion actually is, you risk unconscious behavior. Repressing the emotions is also strongly linked to becoming prone to illness.

ESSENTIAL POINTS: YOUR EMOTIONAL BRAIN

Let feelings come and go. Coming and going are spontaneous.

Don't hold on to negative feelings by justifying why you are right and someone else is wrong.

Look at your emotional weak points. Do you fall in love too easily, lose your temper too fast, become afraid of trivial risks?

Start to observe your weaknesses when they come up.

Ask if you really need to be having the reaction you are having. If the answer is no, the unwanted feelings will begin to go back into balance.

Before the Leap

At this point, we reach a leap in evolution, where the higher brain enters. The question of the meaning of life was born in the cerebral cortex, which sits like a philosopher king atop the lower brain. Kings have been known to topple, and the brain is no exception. The lower brain is always there to make its instinctual, at times primitive,

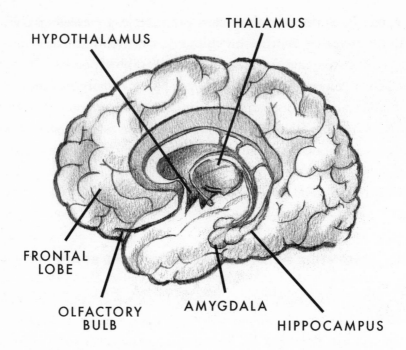

DIAGRAM 3: THE LIMBIC SYSTEM

Tucked under the cerebral cortex is the limbic system (shaded). It houses our emotions, feelings of pleasure associated with eating and sex, and short-term memory. Located here are two individual areas, the thalamus and hypothalamus, as well as the amygdala and hippocampus, which control short-term memory.

The amygdala determines what memories are stored based on the emotional response that an experience invokes. The hippocampus is responsible for short-term memories and sends them to appropriate parts of the cerebral cortex for long-term storage. This region is particularly affected in Alzheimer's disease. The limbic system is tightly connected with the olfactory lobe, which processes smell. This is why a certain scent can trigger such strong memories.

demands. Evolution arguably made no greater leap—either on earth or in the cosmos—than the leap to create the cerebral cortex.

We will give it its own chapter. But first let's look back at the instinctive and emotional brain. They deserve respect for the complexity of their response to the world. If you found yourself being chased by a tiger, the instinctive brain would instantly kick in to release specific neurochemicals that allow you to best survive the chase.

This neurochemical cocktail, primarily composed of adrenaline, required millions of years to be perfected. Adrenaline is the beginning of a chemical cascade in the brain. It evokes electrochemical activity at specific synapses, telling you to run while also optimizing your heart rate and breathing for maximum physical performance. It will also maximize your concentration to endure the chase and outwit the tiger. It will even make you feel pleasure, while subsiding any preexisting feelings of hunger or thirst or even the need to go to the bathroom.

These potential distractions are instantly dissolved so that all physical and mental activity can be focused on escape and survival. When you were in school, if someone challenged you by trying to take away your lunch money, you fought back without thinking about it. Or if the bully was much bigger than you, you fled without thinking about it.

Evolution honed the alliance between the instinctive and the emotional brain to ensure our survival, but if overused, that alliance can become our worst enemy. This is because the instinctive and emotional brains are "reactive"—they mindlessly induce a state of arousal. Any strong external stimulus—a gunshot, the car in front of you suddenly braking, a glance from a pretty girl or flirtatious man—automatically triggers a reaction that triggers the instinct-emotion alliance.

Rudy recalls an experience with bullying in his childhood, one that serves to take us to our next concern, the higher brain. In elementary school he was painfully shy and physically awkward at sports. In contrast, his twin sister Anne was a natural athlete from a young age. When he found himself being picked on by the bullies in the schoolyard, Anne stepped in to fight his battles. It became frustrating that a girl was defending him, and a stronger girl at that.

More important was the frustrated fight-or-flight response, because neither side was succeeding. Running away makes a small child lose his sense of pride; getting beat up is humiliating. Yet in a strange way Rudy was duplicating a primal evolutionary problem. Early humans had to figure out how to live together; they couldn't form a society if they ran away every time adrenaline dictated flight or if they engaged in bloody combat every time adrenaline showed its other side and dictated fight. Rudy had to find a way to solve the same social dilemma. Little by little, as other boys picked on him, he found himself increasingly employing his intellect.

At first the main help was tactical. One time in third grade a bully provoked a fight. The bully hopped onto Rudy's back, pounding away. Anne watched from the sidelines, ready to jump in. But Rudy, instead of panicking and trying to throw him off, had an idea. He noticed a large oak tree behind them and ran backward toward it as fast as he could, pinning the bully against the tree. With the wind knocked out of him, the bully fell off onto the frozen ground and collapsed. With the memory stuck in his mind, that particular boy never bothered Rudy again. In other words, while Rudy's instinctive and emotional brains had warned him of the urgency of the situation, for the first time his intellectual brain had devised a tactic that was neither fight nor flight.

We can imagine early humans making similar discoveries. Once your opponent starts to think, you must do the same. Tactics for fighting war inevitably lead to tactics for ending war. The need to sit by a fire and share the fruits of hunting and gathering leads to reasons for being social. External stimulus wasn't the only prompt that caused the quantum leap in evolution that intellect stands for.

Every cell of the body has an innate intelligence. We cannot limit the far-reaching effect of cellular intelligence, which has been critical to everything that made the body what is it today. Cells live together, cooperate, sense each other, and constantly communicate. If a single cell becomes antisocial and goes rogue, the immune system will intervene, and if that fails, cancer may develop—the ultimate antisocial behavior in the body. In a sense, the higher brain was simply catching up with what every cell knows how to do. Be that as it may, the leap into the intellectual brain increased the possibilities of human life a thousandfold.

SUPER BRAIN SOLUTIONS

PERSONAL CRISES

Many people react to personal crises with fear, which is instinctual. But it's possible to have a more integrated approach, which is to say, using your higher and lower brain together. A personal crisis is just a challenge magnified to drastic proportions, and challenges are part of everyone's life. No one escapes those dark moments when a challenge turns into a crisis; many turning points have come out of impending disaster.

The outcome of your life depends on how you deal with its darkest moments. Will they be turning points or setbacks? What we call wisdom comes into play here, for most people make important decisions based on impulse or its opposite, habit. They feel the tug of emotions, which are never more powerful than when the mind is in disarray. There is no denying the famous first sentence of M. Scott Peck's *The Road Less Traveled*: "Life is difficult." But wisdom can be an incentive to conquer the difficulties, transforming frustration and defeat into turning points and breakthroughs.

Anytime things go badly wrong, ask yourself three questions, all of which are geared to turn the mind's disarray into an orderly process that the brain can follow and organize physically.

DO ASK

1. Is this a problem I should fix, put up with, or walk away from?
2. Whom can I consult who has solved the same problem successfully?
3. How can I reach deeper into myself for solutions?

Conversely, there are three questions that you shouldn't dwell upon because they are self-defeating and promote mental chaos.

DON'T ASK

1. What's wrong with me?
2. Whom can I blame?
3. What's the worst-case scenario?

The situations in which these questions come into play are innumerable, from a bad relationship to a serious car accident, from a diagnosis of life-threatening illness to a child's arrest for drugs. The sad truth is that millions of people constantly dwell on the questions they shouldn't be asking, while only a fraction seriously ask the right questions, leading to the right actions. Let's see if we can improve on that.

1. Is this a problem I should fix, put up with, or walk away from?

The first thing to do is to get your bearings in a reasonable way. Therefore, ask *Is this a problem I should fix, put up with, or walk away from?* Unless you can answer this question clearly and rationally, your vision will be clouded by emotional reactions. Without knowing it, you will be under the sway of the instinctive-emotional alliance in your brain. You may give in to impulsiveness or else fall back on old habits when what you need is something new, a solution that fits the crisis at hand.

Bad situations can often induce bad decision making, and so to get to the point of making good decisions, you must clarify your inner confusion. Pause to consider—with consultation from those you trust—a course of action that begins with finding a fix. If the fix isn't there, ask why. The answer may be that you need to be patient and put up with the bad situation, or else that you need to walk away because no one in your place can find a fix. Financial problems can sometimes be fixed, but sometimes you have to put up with them, unless worse comes to worst and you must walk away by filing bankruptcy. Notice that this sequence has to exist. Society was backward when debt was turned into a moral failing and debtors were thrown into prison. They were deprived of the means to either fix their situation or walk away from it.

Don't trap yourself through judgment and punitive moral attitudes. In general, because finding a fix takes effort and walking away feels risky, most people put up with bad situations, even ones in crisis, such as a violently abusive spouse or serious signs of heart disease due to obesity. Only a small percentage of people (under 25 percent) seek professional help for their emotional problems, while most people (more than 70 percent) report that they deal with emotional difficulties by watching more television.

The alternatives would work if people didn't vacillate when things go bad. One day they wishfully hope for a fix and maybe take a few steps toward it. The next day they feel passive and victimized, so they put up with things as they are. The third day they are sick and tired of suffering and simply want to escape. The overall result is self-defeat. No solution can ever be found by running in three different directions. So clarify your situation and act on what you clearly see.

Actions: When you feel calmer, sit down and examine the crisis. Write down the alternatives, making a column each for *fix it, put up with it,* and *walk away.* Write down the reasons for each. Weigh them carefully. Ask someone you trust to read your list and comment. Once you've decided what to do, stick with it unless strong indications point in a new direction.

2. Who can I consult who has solved the same problem successfully?

Bad situations aren't solved in isolation, but our emotional reactions undoubtedly isolate us. We become afraid and depressed. We draw

into ourselves. Around the edges we entertain shame and guilt, and
once these corrosive feelings take hold, we have even more reason
to shut down. Therefore, you should ask, *Who can I consult who has
solved the same problem successfully?*

Finding someone who has gone through the same crisis that you
are facing accomplishes several things at once. It gives you an example to follow, a confidante who understands your plight, and an
alternative to withdrawing into isolation. Victims always feel alone
and helpless. So reach out to someone who has proven, through their
own life, that you don't have to be victimized by the bad thing you
are facing now.

We aren't talking about hand-holding, shared misery, or even
therapy. All those activities can be beneficial (or not), but there's
no substitute for talking to a person who has entered a dark place
and come out successfully. Where do you find such a person? Ask
around. When you are feeling overburdened and stressed, more
people want to help than you imagine. The Internet widens your
search much further, since it offers active forums, where crises can
be discussed in real time, and links to interconnected sources. But
make sure that you are not entering into a moaning session, either
online or face to face. In the intensity of our feelings, it's easy to lean
on anyone who will give a listen.

Stop and stand back. Are you getting the right feedback? Is
something positive, something you can use, coming out of every encounter? Is the other person truly sympathetic? (You can see through
faking if you allow yourself to.) Sharing our emotions is only the
beginning. You need signs that your emotions are healing and that a
real solution to the crisis is starting to appear.

Actions: Find a confidante to tell your story to. Seek a support
group; go online to find blogs and forums—the possibilities are
much greater than ever before. Don't stop until you find not just
good advice, but real empathy from someone you trust. Put their
words to the test by writing down the solution being suggested. Up-

date these notes every few days until the solution starts to work; otherwise, go back and ask for better advice.

3. How can I reach deeper into myself for solutions?

Finally, there is no getting around facing the crisis head on. Turning a bad thing into a good thing is up to you. No one can be there all the time, and like it or not, crises are all-consuming. You find yourself facing an inner world that is suddenly full of threats, fears, illusions, wishful thinking, denial, distractions, and conflict. The world "out there" won't shift until the world "in here" does. Therefore, ask, *How can I reach deeper into myself for solutions?*

You are seeking entry into the domain of the higher brain, where intellect and intuition can aid you. But you must give permission first, which means a willingness to go deeper inside. We haven't dealt yet with the higher brain. As a preview, consider a simple truth that Rudy and Deepak deeply believe in: The level of the solution is never the level of the problem. Knowing this, you can escape many traps that people fall into.

What exists at the level of the problem? Repetitive thinking that gets nowhere. Old conditioning that keeps applying yesterday's outworn choices. Lots of unproductive, obsessive behavior and stalled action. One could go on. But the relevant insight is that you have more than one level of awareness, and at a deeper level you have untapped creativity and insight.

Your higher brain contains the potential for creating new solutions, but you must cooperate. Many people say, "I have to think this through," which can be a good first step. But at a deeper level the process is one of allowing. You must find a way to hang loose, which is extremely difficult in a crisis. Everyone is tempted to flail. Constant pressure leads to constant worry. Mounting anxiety fuels the lower brain, which amps up its reactions. Only the higher brain is capable of detaching the mind from instinctive-emotional reactions.

So how do you allow the higher brain to function better? Trust and experience both help. If you have, sometime in your past, had *aha!* moments when the solution pops out of nowhere, you can trust that it will occur again. If you value insight, that also helps. Set up the right circumstance for a breakthrough: be quiet for a set part of the day. Close your eyes and follow your breath, until your body begins to calm down. Physical stress blocks the higher brain. Make sure you are well rested, insofar as that is possible. Keep away from stressful triggers and people who make you feel vulnerable.

In your quietness, ask for an answer. For some people, this means praying to God, but it needn't. You can ask your higher self or simply have an intention that is focused and clear. Then back off and relax. Answers always come, because the mind is never at a loss for channels of communication. Putting a question to the universe, as some people would phrase it, stimulates the universe to respond. In any event, generations of wisdom support the notion that creative solutions arise spontaneously.

> The first stage is that fear subsides; you feel strong enough
> to meet the crisis.
> The second stage is that you see what to do.
> The third stage is that you see meaning in the whole
> experience. The higher brain serves this natural
> unfoldment if you allow it to.

Actions: Allow a space for inner quiet. Detach yourself from worry; don't get involved in the chaos. Under these nurturing conditions, you are reaching the level of the solution while detaching from the level of the problem.

The three questions you shouldn't ask will haunt you unless you consciously push them aside. We all feel the urge to condemn ourselves out of guilt, to blame others for our misfortunes, and to fantasize about total disaster. That's what the three bad questions are about, and when we give in to them, they do untold harm in everyday life. Remind yourself, in your moments of clarity, that this is self-punishment. Open a wedge of clear thinking in order to break down the instinctive-emotional reactions that want to take hold.

We can't know exactly what bad things are happening to you. We just urge you to quit being part of the majority who live in confusion and conflict. Join the minority that sees a clear path out of present darkness, that never submits to fear and despair, and that does its part to lead everyone out of crisis into a future full of light.

FROM INTELLECT
TO INTUITION

If the human brain had stopped evolving after the emotional phase, it would still be a marvel. We have extremely subtle emotions that bind us together. But the brain didn't stop there, because the human mind wanted more. It isn't enough to love someone or to feel jealousy, admiration, gratitude, possessiveness, and all the other feelings that are often mixed in with love. It isn't enough that love can be turned up and down, from tender affection to wild passion. The mind wants to dwell on love, to remember who we loved, when, and why. We are the only creatures who could write, "How do I love thee? Let me count the ways." Is it a purely intellectual game? No, it's a way to add a new layer of richness to our lives.

The Intellectual Phase of the Brain

As soon as you ask "Why do I love X?" or "Why do I hate Y?" a more highly evolved element enters—intellect. Intellect is the primary way that your brain has evolved to counter obsessions based on fears and desires. Rational thought allows you to strategize on how to obtain what you desire, an activity that dominates everyone's life. But it also acts as a counterbalance to rein in your emotions. Your emotions and intellect play out their dance at the neurobiological level, as excitatory neurotransmitters like glutamate are engaged in a constant yin and yang with inhibitory neurotransmitters like glycine.

At the level of personal experience, the never-ending interplay between emotion and intellect creates a running internal discourse, which is broadcast in your brain during every waking moment. For some, this discourse takes the form of an internal monologue in which the brain is doing all the "speaking," as it draws from old memories, habits, and conditioning. For others, the discourse is more an internal dialogue, where old and new ideas contend. The person must decide which to favor, the brain's wired-in reactions or new and unknown responses. That can be a problem.

The struggle is difficult enough that some people try to live a life of pure intellect, denying their emotional side. Jesse Livermore was an iconic investor in the stock market during the Roaring Twenties. Born in Massachusetts in 1877, he stares at us blankly and rather dourly today from old photographs. But he was among the first financiers who held no job in his life other than manipulating the numbers on a ticker tape. He lived for numbers and regulated his life with absolute precision. He left home at 8:07 every morning, and at a time when stoplights were hand-controlled by policemen standing on boxes, the sight of his limousine caused every light on Fifth Avenue to turn green.

On October 29, 1929, the disastrous Black Tuesday when the stock market crashed, Livermore's wife assumed that he had lost his fortune, as all their friends had. She ordered the servants to clear the furniture out of their mansion, and Livermore came home to an empty house. But actually he had listened to what the numbers told him and managed to make more money that day than he ever had before. This may seem like a triumphant story for pure intellect, but during the 1930s regulation came to Wall Street. The buccaneering days, when a few rich investors could manipulate stocks at will, were over. Livermore found it hard to adapt. His trading turned erratic.

He became discouraged, then depressed, and in 1940 he retreated into the bathroom of his private club and shot himself in the head. It was never revealed what happened to his millions.

It comes naturally to our intellect to ask questions and look for answers. The human mind has an endless craving for knowledge. We live on two parallel tracks. On one track we experience everything that happens to us, while on the other we question those experiences. The cerebral cortex, the most recent addition to the brain, takes care of thinking in all its aspects, including decision making, judgment, cogitation, and comparisons. To a neurologist, the cortex is the most enigmatic part of the brain. How did neurons learn to think, and even more mysteriously, how did they learn to think about thinking?

For that is what you do every day. You have a thought, and then you reflect on what the thought means. This sounds overly abstract, so let's diagram it from the brain's perspective:

Instinctive: "I'm hungry."
Emotional: "Mm, banana cream pie would taste so good
    right now."
Intellectual: "Can I afford the calories?"

In the intellectual phase, you have endless choices. You can ask yourself, "Who makes a good banana cream pie?" or "Is that what I really want?" or "Does this mean I'm pregnant?" You can think anything you want to, including the most far-out idea ("Do bananas feel pain when they are picked from the tree?"), the most imaginative ("I'd like to write a children's book about a boy who meets a talking banana cream pie"), and everything in between.

We humans are proud of our intellect, to the point that until recently we have denied that lower animals have any kind of intelligence. That's changing quickly. Few birds winter on the snowbound north rim of the Grand Canyon, for example, and some that do spend the autumn months picking seeds to bury in the ground. They harvest pine cones for nuts, and they give each tiny one a burial spot, apparently at random, until hundreds have been deposited. When the winter blizzards arrive, these sites are covered with snow. Yet the birds have been observed going back to each place where a pine nut is buried, pecking through the snow, and retrieving it. Each bird returns only to its own buried food, without poaching at random on the cache placed by other birds.

Examples of animal intelligence are myriad, yet we still remain certain that intellect is exclusively human. Brain structure bears this out, since relative to our brain size, which is very large for our weight, a disproportionate part belongs to the higher brain. (The fact that 90 percent of your cortex is the neocortex, the "new bark," shows that you do a great deal of thinking and deciding, while a dolphin's large brain is about 60 percent devoted to hearing, which makes sense for a creature that is guided by underwater sonar.) Despite the notion that we are driven by lower impulses like sex, hunger, anger, and fear, the higher brain dominates everything. After all, before two countries can go to war and bomb each other's cities, they first must build those cities—and those bombs—which represents a massive accomplishment by the intellect.

The higher brain marks the arrival of self-awareness. Every example we've given uses the word *I* as part of the thought; *I* is the conscious being who is using the brain. The instinctive and emotional phases of the brain dwell in the world of the subconscious. We suppose that animal intelligence is entirely subconscious. On the same phase of the moon in May, horseshoe crabs come ashore by the tens of thousands to lay their eggs on the Atlantic seaboard of North America. They gather from the ocean depths, as they have done for hundreds of millions of years. Within the next few days, a tiny bird known as the red knot sandpiper (*Calidris canutus rufa*) arrives on its migratory route to feed on the horseshoe crab eggs scattered in the sand.

Red knots, small brown-speckled birds that step gingerly on stilted legs, spend the winter in Tierra del Fuego, thousands of miles away in the southern hemisphere, where they feed on tiny clams. No one knows why red knots migrate 9,300 miles between the Antarctic and the Arctic, where they will raise their young. Even less do we know how the red knots learned to time their migration to correspond with the last full moon or new moon in May, exactly when horseshoe crab eggs are lying around the beaches of Delaware Bay, becoming the only food that red knots eat during their stopover. Where the birds are headed, Southampton Island in Canada, is windy, bare, and bleak, affording almost no food. The highly fatty horseshoe crab eggs allow them to store up enough energy to survive. The whole complex setup implies that instinct isn't always simple or primitive. It achieves things that intellect cannot yet fathom.

Is all of nature really unconscious, or are we trapped by our desire to label it that way? One thing is certain. In humans, the intellectual phase of the brain blends instinctive drives and emotions with knowledge gained from experience. If a person's experiences are unhappy, the intellect can try to find better experiences, or it can take more drastic steps to end misery, such as through suicide. It was depressing but insightful for Nietzsche to say, "Man is the only animal who has to be encouraged to live." There's a more positive way to say the same thing: humans refuse to be dictated to by our lower brain, even when it comes to survival.

The intellectual brain uses logic and rational thought in order to deal with the world in a mindful manner. While the instinctive brain causes you to naturally and innately *react*, the intellectual brain provides you with the option of mindfully *responding*. Response comes from the Latin root *responsum* and refers to reacting in a *responsible* manner. Responding to any situation requires understanding, while reacting doesn't. Understanding isn't an isolated event. There's always a social context. You must empathize with others; people must communicate and make meaningful connections. Conceivably,

Homo sapiens could have remained sociable without these higher traits. Chimpanzees are sociable, and they broke off the primate family tree six million years after, not before, our hominid ancestors.

Looking into a chimpanzee's eyes, one detects moments when the animal seems thoughtful, but chimps are not responsible, and for all their intelligence, they cannot push their learning curve. You can set up an experiment in which a chimpanzee watches while you hide some food under one of two boxes. If he remembers and looks under the correct box, he gets the food. It takes only a few tries for chimpanzees to learn to succeed at this. However, let's say you change the experiment. You place two boxes in front of the chimpanzee, and if he hands you the heavier box, you give him a food reward. Even after six hundred tries, a chimpanzee will not perform better than randomly on this test. A young child of three or four figures out very quickly that it needs to choose the heavier box.

We also share our learning. Human society depends on teaching, which requires a special kind of brain, one that instantly turns experience into knowledge. After millions of years, some monkeys have learned to smash hard nuts on rocks to break them open, and higher primates like chimpanzees can use a stick to pry bird's eggs out of deep holes in a tree trunk or ants from a hole. But this skill remains primitive. An orangutan can be taught to retrieve food from a complicated plastic container that has several moving parts that must be opened in a precise sequence. Orangutans are quick at solving this puzzle, but then they run into a block: they can't teach another orangutan how to solve the same puzzle.

We don't teach just by example, either, but by talking. Complex language accelerated the evolution of the brain, because it allowed for a more sophisticated mode of communication. It also allowed us to be capable of symbolic thought. This means we can create symbolic or virtual worlds using the same parts of the brain that evolved for communicating with one another. When you stop at a red light, you aren't stopping because you hear the word *stop*. Rather, you connect the color red with the word; it's a symbol. Simple as that sounds, it has enormous ramifications. Dyslexic children, for example, have learning difficulties with reading due to a defect in brain development in the womb. Their brains put words and letters in reverse order. However, it has been discovered that this defect can be bypassed by using colored letters of the alphabet. *A* might be red, *B* green, and so on. With this symbolic association, language can proceed because a brain mechanism in the visual cortex has been appropriated for a new use: the ability to distinguish colors, which in humans extends to incredible subtlety: the human eye can detect 10 million different wavelengths of light. No one knows exactly how many of these translate into colors that we can discriminate between, but there seem to be several million at least.

This tremendous gift of imagination and symbol making can be turned against itself. The swastika originated as an ancient Indian symbol for the sun, but if painted on the side of a synagogue, it denotes desecration or even a hate crime. Image can also block reality. The phrase *movie goddess* was invented to reinforce the public's fantasy that Hollywood actors aren't like regular people. As a result, however, the public craves a peek behind the image, and the more tawdry and sordid the reality being exposed, the more titillating it is.

There's a long history of dividing the mind into instinct, intellect, and emotions. Neuroscience can now map the regions of the brain corresponding to each. But it's worth remembering that these divisions are only models that were invented because Nature is so hard to grasp in its full complexity. In truth, we are constantly making reality, a process that embraces every region of the brain in a constantly shifting interplay.

As with any phase of the brain, intellect can go out of balance.

If you are too intellectual, you lose the grounding of emotions and instincts. This leads to overly calculated actions and castles in the air.

If you don't develop your intellect, it remains stuck in rudimentary thinking. This leads to superstition and falling prey to all kinds of faulty arguments. You become the pawn of influences from outside yourself.

ESSENTIAL POINTS: YOUR INTELLECTUAL BRAIN

Intellect stands for the mind's most recent evolutionary phase.

Intellect never operates in isolation but is blended with emotions and instinct.

Intellect helps you to rationally deal with your fears and desires.

Responding to the world implies being responsible for the world.

Rational thought becomes destructive when it forgets its responsibilities. (Hence the rise of atomic weapons, the destruction of the ecosystem, etc.)

The Intuitive Phase of the Brain

Your intellect is part of your birthright, which includes an insatiable need for meaning. You inherited intuition out of a different need that is just as powerful: the need for values. Right and wrong, good and bad, are so basic that the brain is wired for them. From a very early age, infants seem to display intuitive behavior in this department. Even before the toddling stage, a baby who sees his mother drop something will offer to pick it up for her—helping is a built-in response. A two-year-old can be shown a puppet play in which one puppet does nice things while the other does the opposite. Nice

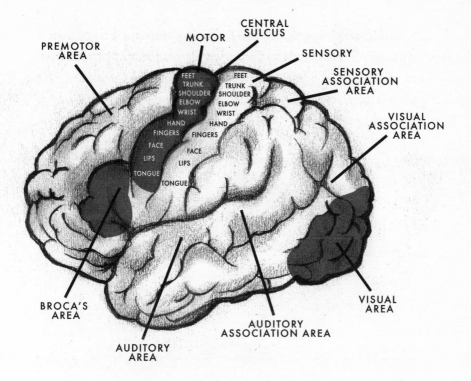

**DIAGRAM 4: THE FUNCTIONAL AREAS
OF THE CEREBRAL CORTEX**

The bulk of the brain is the cerebral cortex or cerebrum. Designated the higher brain, it is responsible for many functions we associate with being human: receiving and processing sensory information, learning, memory, and the initiation of thought and action, as well as behavior and social integration.

The cerebral cortex is the most recently evolved part of the brain, consisting of a roughly three-square-foot sheet of neural tissue spread out in six layers toward the outer surface of the brain. (*Cortex* means "bark" or "rind" in Latin.) This sheet of tissue is folded upon itself many times over so that it can fit into the skull. The cerebrum is home to the largest concentration of neurons in the entire brain, roughly 40 billion.

The cerebral cortex has three main functional areas: the sensory regions for receiving and processing the five senses, the motor regions for controlling voluntary movement, and the association regions for intellect, perception, learning, memory, and higher order thinking.

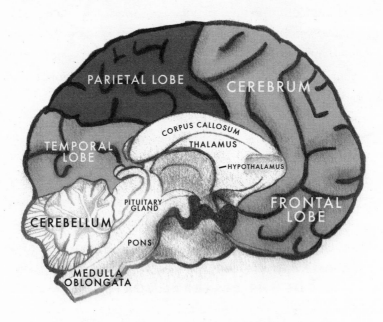

DIAGRAM 5: THE REGIONS OF THE CEREBRAL CORTEX

The cerebral cortex is made up of a number of different lobes. Toward the back of the cerebral cortex is the occipital lobe, containing the visual cortex, where the brain relays and interprets information being perceived by your eyes. The left visual cortex connects with the right eye and vice versa. Toward the front of the occipital lobes are the temporal lobes. Here lie the primitive instinct-driven emotions that serve survival: fear, desire, and appetites such as hunger and sexuality. Hearing and balance are also controlled here. When this area of the brain is damaged or malfunctioning, a person may suffer uncontrollable appetites for food and sex.

In front of and above the temporal lobes are the parietal lobes, where sensory information is processed along with spatial orientation, which lets you know where you are. Finally, in front of the parietal lobes are the frontal lobes. The frontal lobes regulate motor control and movement but also mediate our behavior in society. If the frontal cortex is damaged or, for example, contains a tumor, one might become pathologically uninhibited and turn into an extreme exhibitionist or even a sexual molester.

The right and left hemispheres of the cerebral cortex are connected by bundles of nerve fibers called the corpus callosum. They allow the two sides of the brain to "talk" to each other. If they did not, one might experience "alien hand syndrome," in which one does not recognize one's own hand! Tucked under the corpus callosum is the limbic system (see Diagram 5), which contains the thalamus and hypothalamus. The thalamus is involved in sensory perception and regulates movement. The hypothalamus regulates hormones, the pituitary gland, body temperature, the adrenal glands, and many other activities.

The two other major sections of the brain are the cerebellum, toward the rear of the brain, which controls coordination of movement, balance, and posture; and the brain stem (medulla oblongata and pons), which is the oldest part of the brain. It connects the brain to the spinal cord and regulates heart rate, breathing, and other so-called autonomic processes that take place automatically.

The functions of the brain that control physiological processes—from heart rate to fear response to the immune system—are concentrated in specific regions of the cerebral cortex, cerebellum, or brain stem. But these regions also communicate with one another to create an intricate system of balance and coordination as part of every brain activity. For example,

when you look at a flower, your eyes sense that visual information and relay it to the occipital cortex, a region of the cerebral
cortex toward the back of the brain. But first that same visual
information travels through multiple other areas of the brain,
where it may also serve to coordinate your movements in response to the visual information. The billions of neurons in these
regions work together in exquisite balance and harmony, similar
to the way an orchestra makes beautiful music. There is no room
for one instrument to be too loud or off key. Balance and harmony are the keys to a successful brain, just as they are for the
stability of the universe.

involves playing and cooperating; not nice involves being selfish and
complaining. When asked which puppet it likes best, a child will
pick the "good" puppet much more often than the "bad" one. We
have evolved with brain responses for morality.

But intuition has also been a suspect area. A curious irony of the
brain is that the intellectual brain can dismiss the intuitive brain as
mere superstition, bordering on belief in the paranormal. Rupert
Sheldrake, a far-seeing British biologist, has done decades of experiments to verify intuition. For example, he has tested the common
experience of feeling that you are being watched, usually by someone
standing behind you. Do we have eyes in the back of our head? If
so, this would be an intuitive ability, and Sheldrake has shown that
it exists. For his pains, his work is considered controversial, which
means, as Sheldrake wryly notes, that skeptics haven't actually bothered to look at his results.

But the fact that humans are intuitive is uncontroversial. Whole
areas of your life depend upon intuition—empathy, for example.
When you walk into a room, you can sense if the people in it are
tense or have been fighting before you arrived—that's intuitive. You
intuit when someone is saying A but means B, or when someone
who is holier than thou is hiding a secret.

Empathy is defined as the understanding and sharing of others' feelings. In *Homo sapiens*, as the ability to communicate took a quantum leap forward, empathy became a critical component for social survival. It allowed parents to care for the group's children while some adults were absent to hunt and gather. Empathy still enables us to live in groups and socialize with each other, serving as the necessary curb to selfish aggression and competition (a balance that society struggles to maintain).

More broadly, empathy has paved the way for moral reasoning and altruistic behavior. (The Latin roots of *compassion* mean "to suffer with," pointing to our ability to mirror what we see another person feeling.) Empathy is different from sympathy, which does not involve sharing another's state of mind. It is also different from *emotional contagion*, in which one is not aware of whether the emotion belongs to oneself or has been absorbed by contact with a stronger personality, or the crowd.

At the neural level, the main area of the brain activated by empathy is the cingulate cortex. *Cingulum* means "belt" in Latin. The cingulate lies like a belt in the middle of the cerebral cortex and is considered part of the limbic system, which deals with emotion, learning, and memory. This is where empathy physically resides. The empathy-associated regions of the cingulate gyrus are larger in females than in males and are generally smaller in schizophrenic patients, who are often tragically isolated in their emotions and delusional about what other people are feeling.

Empathy has also been associated with *mirror neurons*, a class of nerve cells that are known to exist in lower primates like monkeys. There is a neuronal reason for "monkey see, monkey do" that is critical to learning new skills. When a baby monkey, even one still young enough to breast-feed, sees its mother grasp food and eat it, the areas in its brain responsible for grasping, tearing food apart, and chewing light up—they mirror what it sees. Experiments cannot be performed on human infants to determine if the same is occurring in our species, but in all likelihood it is. (The pernicious side of mirroring may be that when a young child witnesses negative behavior, such as domestic abuse, a brain pattern may be triggered. It is known that abused children often grow up to become abusers, so imprinted are they with such behavior.)

No one knows the full functioning of mirror neurons, but they seem to play a key role in social attachment, the process by which we attain security, nurturing, and alleviation of distress from our relationships. A host of neurochemicals called neuropeptides—small proteins in the brain that regulate social attachment, including oxytocin, opioids, and prolactin—regulate empathic responsiveness.

Oxytocin facilitates maternal behavior and makes one feel "in love." The application of oxytocin via nasal spray has been shown to reduce responses to social stress and the fear responses in the brain. Oxytocin can also increase mutual trust and make one more sensitive to others' facial expressions. An adverse gene mutation in the receptor that binds oxytocin causes one to have lower levels of empathy. Thus oxytocin plays a critical role, and yet its popular name, "the love hormone," should not be taken literally. Love, being a complex behavior, is sensitive to many responses throughout the brain, and a single hormone cannot be held as its cause. We are confronted with the riddle of where mind ends and brain begins. Anyone who has ever fallen madly in love will testify that this mystery gets very personal. Humans have evolved a biological structure in the brain that grew from the mating of lower mammals, but we make all kinds of choices about how we love and who attracts us. Biology may provide the juice, but it doesn't take over from the mind.

All such issues lead back to free will, which we believe is always dominant in human life. But we can interpret the fact that neurochemicals can control our emotions, including love and empathy, in two ways. On one hand, we can say we have no control over how we feel; we are slaves to our neurochemistry with little or no free will. Alternatively, from the super brain point of view, we can argue that the brain is an incredibly fine-tuned organ that produces the emotions we need at any given moment. The brain needs triggers, which can be very subtle. Meeting an attractive man is different for a woman depending on whether she is "in the market." If she is not, her brain's love mechanism isn't triggered; if she is, the opposite is true. In either event, the brain did not make the woman's decision for her. Despite their undeniable power, our emotions are generated to serve us.

This is where intuitive mind enters the picture. It rises above emotion and intellect, giving you an overall picture of things (which psychologists call a *gestalt*, the image of reality we assign to various situations). At work, the person in charge doesn't have to wear a sign that says, "I'm the boss." All kinds of signals (such as his tone of voice, his big office, his air of authority) merge into a picture that we intuitively grasp. We say that we "feel out" a situation, but this isn't the same as an emotion. It's the feeling that tells you what is going on all at once, instead of having to assemble a picture one emotional or intellectual piece at a time.

All of the following things fall under the category of intuition:
Falling in love at first sight.
Knowing that someone else is lying.
Feeling that things happen for a reason, even if the reason hasn't yet emerged.
Using irony, which says one thing but means the opposite.
Laughing at a joke.

Intuition would be less controversial if it were isolated in a specific location in the brain, but it isn't. The most popular belief is that the right hemisphere of the brain is responsible for intuition while the left is rational and objective, but this neat division hasn't held up to rigorous testing. Still, the hallmarks of intuitive people are well confirmed:

They make quick decisions without going through a ratio-
    nal process, yet their decisions are just as accurate.
They pick up on subtle facial expressions.
They rely on insight, defined as knowing something
    directly without waiting for reason to arrive at a
    conclusion.
They make creative leaps.
They are good judges of character—they know how to read
    people.
They trust and follow their first instinct, in so-called
    "blink" or snap judgments.

For anyone who trusts their intuition, this last category of snap
judgments is especially intriguing. Traditionally, we value other
kinds of judgments more. Young people are advised not to be rash,
to think things through, and to arrive at a considered judgment.
But in reality we all make snap decisions. Hence the saying that
you can't take back a first impression. First impressions, made in
the blink of an eye, are the most powerful. What has emerged from
recent studies is that first impressions and snap judgments are often
the most accurate. Experienced real estate brokers will tell you that
home buyers know within thirty seconds of walking into a house
whether it is right for them or not.

It was long assumed that a person can recognize faces better if
he first goes through the process of verbally describing the face. *The
girl had long brown hair, fair skin, a button nose, and small blue eyes* is
supposed to help fix a certain face in your memory. But experiments
show the opposite. One study flashed a series of photos in fast succession, asking subjects to press a button if they saw a particular face
flash by. People who had glimpsed the face only briefly did better at
this than people who saw the face and were given time to verbalize
its features. Such findings seem intuitively right (there's that word
again), because we all know what it means to have someone's face

stick in the mind even though we don't rationally break it down into separate features. We also believe victims of a crime who say, "I'd know that face in a million years if I ever saw it again."

In effect, intuition fits the bill for anyone seeking a sixth sense. A sense is basic, a primal way to take in the surrounding world by looking, listening, and touching. More important, you "feel" your way through life, following hunches, knowing what's good for you and what isn't, where you should aim your career and avoid a dead end, who will love you for decades and who is only a passing fancy. Highly successful people, when asked how they reached the top, tend to agree on two things: they were very lucky, and they wound up in the right place at the right time. Few can explain what it takes to be in the right place at the right time. But if we value intuition as a real skill, highly successful people are probably the best at feeling their way through life.

Seeing the future is intuitive, too, and we are all designed for it. There is no need to call the ability paranormal. In one experiment, subjects were shown a fast series of photos, some of which were horrendous depictions of fatal auto crashes or bloody carnage in war. The subjects were monitored for signs of stress response, such as faster heart rate, rising blood pressure, and sweaty palms. As soon as a horrifying image was presented, it inevitably triggered the stress response. Then a strange thing occurred. Their bodies began to indicate stress just *before* a shocking image was displayed. Even though the photos were shown at random, these people reacted in anticipation of being shocked; they didn't react in anticipation of innocuous images. This means that their bodies were predicting the future, or to be more precise, their brains were, since only the brain can trigger the stress response.

We aren't promoting one phase of the brain over another. But it's crucial not to deny a phase out of stubborn skepticism or intellectual bias. Controlled studies are meant to be objective proof of the kind that the intellect will accept, so it's unfair for hundreds of studies in cognitive psychology to prove that intuition is real, while our social attitude toward intuition is largely doubtful and even negative. Are you intuitive? Your intuition tells you that you are.

As with any phase of the brain, intuition can go out of balance.

If you are too trusting of your intuitive hunches, you fail to see reason when it counts. This leads to impulsive decisions and irrational behavior.

If you ignore your intuition, you lose the ability to feel out situations. This leads to blind decision making that depends too much on rationalizing your actions, even when they are obviously wrong.

ESSENTIAL POINTS: YOUR INTUITIVE BRAIN

Intuition can be trusted.

"Feeling" your way through life brings good results.

Snap judgments are accurate because intuition doesn't need processing by the higher brain.

Reason is slower than intuition, but we often use it to justify intuition, because we have been taught that reason is superior.

The intuitive brain has no limits that are foreseeable—everything depends on what the mind wants the brain to do.

Putting It All Together

Having taken apart the fourfold brain, what do we get once it is put back together? A superb tool for reality making, which has infinite possibilities. The best way to achieve health, happiness, and success is by balancing all four phases of your brain. Your brain goes out of balance when you favor one part over another. Notice how easy it is to identify with one phase of the brain, which encourages it to dominate. If you say, "I'm sad all the time," you are identifying with the emotional brain. If you say, "I've always been smart," you are identifying with the intellectual brain. In the same way, you can be dominated by the instinctive brain when you are obeying unconscious urges or the intuitive brain when you follow hunches, gamble, and take risks. With enough repetition, the favored regions of the brain gain an advantage; the unfavored regions start to atrophy.

But your true identity isn't found in any of these separate regions. You are the summation of them all, as the mind controls them. The shorthand for the brain's controller is *I*, the self. *I* can forget its role and fall prey to moods, beliefs, drives, and so on. When this happens, your brain is using you, not out of spite or in order to grab power. You have trained it to use you. It's hard to really absorb that every thought is an instruction, but it is. If you stop in front of an Impressionist painting, the brilliant colors and airy mood are instantly appealing. None of the raw data being processed by the visual cortex is training the brain. (You mastered the basic skill of focusing your eyes on a specific spot without wavering in the first months of life.) But as soon as you think, *I love this Monet cathedral,* you are instructing your brain—in other words, training it—and not in a simple way.

At the instant you think *I like X,* whether it is Monet, a banana split, or the person you will marry one day, your brain goes into holistic mode.

It remembers what you like.
It registers pleasure.
It remembers where the pleasure came from.
It makes a note to repeat the same pleasure in the future.
It adds a unique memory to your memory bank.
It compares the new memory to all previous ones.
It sends chemical reactions of pleasure to every cell in your
   body.

This is actually only a brief sketch of what it means for your brain to kick into holistic mode. It would be tiring to describe every detail, but at the very least you know what museum you are standing in, how people are moving around the room, and whether you feel tired or not, along with the usual unconscious things like feeling hungry or wondering if your feet are hurting from walking too much.

Putting it all together is the human mind's single greatest achievement. It is what we do, yet by no means can we explain how it is done. Experience is infinitely richer than explanations. Our aim is to expand the brain's holistic mode. Deep down we all know that it is better to love all the paintings in a museum, not just a select few. Each painter has a unique vision, and when you appreciate the art, you have opened yourself to the vision. Even deeper down we know that it is better to love all people than to love just the few who are closest to us. But to expand the brain's emotional centers is threatening. We typically identify with people who are most like us (in race, status, education, politics, etc.) and feel alienated by those who are most unlike us.

As you grow older, your tendency is to narrow your likes and dislikes, which means that you are denying your brain its ability to be holistic. An interesting experiment in social psychology took ten people from Boulder, Colorado, a very liberal town politically, and ten people from Colorado Springs, which is traditionally very conservative. One of the problems with present-day America is its divisive politics, and there's a demographic reason behind it. In the past people who had opposite political views lived together, and therefore a candidate running for office might win by five or six percentage points.

Since World War II, however, there has been a decisive shift. Liberals have moved to towns where other liberals live, and conservatives to towns where other conservatives live. As a result, elections are now grossly one-sided, and candidates typically win by large margins. The Boulder–Colorado Springs experiment wanted to test if this can be changed. Sitting in their own groups, the ten subjects from each town discussed politics and rated themselves on how they felt on each issue. For example, when it came to abortion or gay marriage, they marked their position from one to ten depending on how extremely they stood for or against.

At this point one person from Boulder sat in on the Colorado Springs group and one person from Colorado Springs sat in on the Boulder group. Each was allowed to argue for his stance, pleading for liberalism or conservatism to people who leaned the other way. After an hour, the groups again rated themselves on hot-button issues. Did hearing from the other side soften their opinions? You might think so, but in fact the opposite happened. After hearing from a liberal, the conservatives became *more* conservative on the issues than they had been before. Likewise, the liberal group became more liberal.

Assessing these findings can lead to discouragement. One would like to think that being exposed to other viewpoints opens the mind. But some neuroscientists conclude from such findings that us-versus-them thinking is wired into the brain. We define ourselves by opposition; we need enemies in order to survive, since by having enemies early humans sharpened their skills at self-defense and warfare.

We are strongly opposed to such interpretations. They ignore a basic fact, that the mind can overcome the wired-in pattern of the brain. In the case of the Boulder–Colorado Springs experiment, there's a huge difference between listening with a closed mind to an opposite opinion and deciding that you *want* to understand it.

A funny-sad story comes from a friend of Deepak's who was born in the South. His small North Carolina town had a department store named Bernstein's, which was Jewish. It had other families, not Jewish, who were also named Bernstein. "The non-Jewish ones pronounced their name Bern-stine, while the department store was Bern-steen," Deepak was told. Why? His friend shrugged. "That was the only way anybody knew who to be prejudiced against. To tell the truth, nobody in my family had actually ever met anyone who was Jewish."

We refuse to believe that the tendency to discriminate comes from hardwiring. If you examine its physical design, the brain is a highly integrated organ in which various regions and their resident nerve cells are constantly communicating with each other. To a biologist, all traits, including the brain's ability to communicate among billions of neurons, can be reduced to two main goals: survival of the species and survival of the individual. But present-day human beings don't accept mere survival. If we did, there would be no charity for the poor, hospitals for the sick, and care for the disabled.

Preserving everyone's life, not simply the alpha types who can get the most food and mating rights, lifted us above Darwinian evolution. We share food; we can marry without having children. In short, we are evolving as a choice, not as nature's necessity. The brain is moving in a more holistic direction.

Our favorite phrase for this trend is "survival of the wisest." If you choose to, you can evolve through conscious choices.

WHERE THE BRAIN IS GROWING
HOW TO BECOME PART OF
THE NEXT EVOLUTIONARY LEAP

Don't promote conflict in any area of your life.
Make peace when you can. When you can't, walk away.
Value compassion.
Choose empathy over blame or derision.
Try not to always feel that you are right.

Make a friend who is the opposite of you.

Be generous of spirit.

Wean yourself off materialism in favor of inner fulfillment.

Perform one act of service every day—there is always
something you can give.

Show genuine concern when someone else is in trouble.
Don't ignore signs of unhappiness.

Oppose us-versus-them thinking.

If you are in business, practice capitalism with a conscience,
giving ethical concerns as much weight as profits.

These are not merely ideals. Dr. Jonas Salk, who gained world fame for the cure of polio, was also a visionary and philanthropist. He developed the concept of the "metabiological world," a world that has gone beyond biology. Such a world depends upon human beings in our role as reality makers: everything we do, say, and think goes beyond biology. But what is the purpose of everything we do, say, and think? For Salk, we have a single overriding purpose: to unfold our full potential. Only the holistic brain can get us there. On its own, science, being intellectual, excludes the subjective world of feelings, instincts, and intuitions. To most physicists, the universe has no purpose; it's a vast machine whose working parts exist to be figured out. But if you use your whole brain, the universe certainly does have a purpose, to foster life and the experiences that life brings. When your own experiences become richer, the universe gets better at serving its purpose. That's the reason why the brain began to evolve in the first place.

SUPER BRAIN SOLUTIONS

FINDING YOUR POWER

If everyone has the power to make reality, why do countless people live in dissatisfaction? Reality making should lead to a reality you actually want, not the one you find yourself in. But that can't happen until you find your power. As with everything else, personal power must go through the brain. A powerful person is the combination of many traits, each of them trained into the brain:

WHAT'S IN PERSONAL POWER?

Self-confidence
Good decision making
Trust in gut feelings
Optimistic outlook
Influence over others
High self-esteem
Ability to turn desires into actions
Ability to overcome obstacles

Whenever someone feels powerless to change a situation, whatever it may be, one or more of these elements is missing. You may imagine that powerful people are born with an extra dash of confidence and charisma. But the most powerful CEOs tend to be quiet, organized people who have learned the secret of shaping situations toward goals they want to achieve. Each started at a point not very different from anybody else. The difference has to do with feedback. They internalized every small success and reinforced the next opportunity. They trained the brain by absorbing experience and pushing the bar higher.

People who feel powerless, on the other hand, have trained themselves by absorbing negative experiences. The process remains the same so far as the brain is concerned. Neurons are neutral about messages of success or failure. In an ideal world, the title of this section would be "Five Ways to Feel More Powerful." But as things stand, many people feel powerless, and the social trends that drain personal power only grow stronger. Whether you struggle due to the recession, a controlling spouse, or the anonymity of routine work, it's crucial to find your power, all the more so when the world's wisdom traditions keep repeating, age after age, that infinite power is hidden inside every individual.

We'd like to be systematic and clear up some basic mistakes. Before talking about personal power, let's clarify what it isn't. It isn't force that you use like a weapon to get your own way. It isn't suppressing what you don't like about yourself and achieving a perfect ideal for the world to admire. It isn't money, status, possessions, or any other material surrogate. There are heirs to fortunes, sitting in the lap of luxury, who feel more powerless than the average person. This is so because the issues of power are all "in here," where you relate to yourself.

Now that we know what personal power isn't, we can list the five steps that bring true power.

1. Stop giving away your power.
2. Examine why it's "good" to be a victim.
3. Develop a mature self.
4. Align yourself with the flow of evolution, or personal growth.
5. Trust in a higher power that transcends everyday reality.

Each of these points depends on a single thread that ties them all together: the reality you see all around you has been constructed from invisible currents flowing in, around, and through you.

"In here" you are supported by the creativity and intelligence of your body, with its innate wisdom. "Out there" you are supported by the evolutionary force that sustains the universe. To believe that you are disconnected from these powers, sitting alone and weak in a private bubble, is the fundamental mistake that leads to feeling powerless in everyday life.

Let's look at each step for reconnecting to the source of personal power.

1. Stop giving away your power.

Becoming powerless doesn't happen in a single dramatic stroke, like the barbarian hordes breaking down your door and burning your house. It's a process, and for most people, the process is so gradual that they don't notice it. They are more than happy, in fact, to give away their power by degrees. Why? Because being powerless seems like an easy way to be popular, accepted, and protected.

You are giving away your power when you please others in order to fit in.

Or when you follow the opinions of the crowd.

Or when you decide that others matter more than you do.

Or when you let someone who seems to have more power take charge of you.

Or when you hold a grudge.

All of these actions occur at the psychological level, which is invisible. If a woman gives away her power without noticing, it then seems only right and proper for her to be modestly sitting in the background, holding accepted opinions, living for the children, and letting a controlling spouse run roughshod over her in order to keep the peace. In small and large ways, such sacrifices reduce her sense of self-worth, however, and without self-worth, she diminishes what her brain can do by giving it low expectations.

All hidden power is self power. If you chip away at your self-worth, what replaces it is a series of compromises, false gestures, habits, and conditioning. Your brain gets trained to view life as a gradual decay in exciting challenges, and without such challenges, reality making becomes a routine affair. Low self-esteem serves as a filter that blocks the signals being constantly sent to you for being successful.

Breaking out: To stop giving away your power, resist the urge to go along. Learn to speak up for yourself. Stop postponing the little things you dread doing. Give yourself a chance for a small success every day. Notice your successes, and let them register as moments of fulfillment. Stop equating self-denial with virtue. Getting less so that others get more is a rationale for lack of satisfaction. Stop holding grudges and wasting energy on sustained anger. The next time you feel a threat, ask how you can turn it into an opportunity.

2. Examine why it's "good" to be a victim.

Once you start chipping away at your self-worth, it's only a short step to victimization. We define being a victim as "selfless pain." By saying that you don't really count, you can make the suffering you endure into a kind of virtue, as all martyrs do. It's good to be a martyr when you serve a higher spiritual purpose—or so some religions believe—but what if there is no higher purpose? Most victims sacrifice themselves on the altar of worthless causes.

"GOOD" SUFFERING YOU DON'T NEED

Taking the blame for someone else's mistakes.

Covering up abuse, physical or mental.

Allowing yourself to be belittled in public.

Letting your children disrespect you.

Not speaking your truth.

Denying yourself sexual fulfillment.

Pretending to love.

Working at a job you hate.

To indulge in even one of these pointless kinds of suffering makes you far more vulnerable to bad things in general, since victimization, once it becomes a habit in the brain, restricts your responses. Unconsciously, you decide in every situation that you are chosen to take the brunt of trouble. That's a very dangerous and powerful expectation.

Victims always find "good" reasons for their plight. If they are forgiving an abusive spouse, forgiveness is spiritual, right? If they are enabling an addict, tolerance and acceptance of others is equally spiritual. But if you stand back, victims in such situations are deliberately bringing suffering upon themselves, and ultimately it comes down to powerlessness. A victim is always being done to. There are enough abusers, addicts, rage-aholics, control freaks, and petty tyrants to drain the power from anyone who volunteers to play the role of victim.

Breaking out: First and most important, realize that your role is voluntary. You are not trapped by fate, destiny, or the will of God. The whole mindset that "good" suffering is holy may be true for saints, but in everyday life, to remain a victim is a bad choice. Turn your choices around. Recognize who you have hired to be your victimizer, and take steps to fire them. Don't procrastinate, and don't rationalize. If you feel abused, hurt, belittled, or done to in any way, face the truth and get out as quickly as possible.

3. Develop a mature self.

Human beings are the only creatures who do not mature automatically. The world is full of people stuck in childhood and adolescence, no matter how old they happen to be. To mature is a choice; to reach adulthood is an achievement. Bombarded by mass media, it's easy to mistake youth for the prime of life, when in fact the young (from around thirteen to twenty-two) are going through the most troubling, insecure, and stressful phase of life. No project is more decisive for personal power—and happiness—than the project of becoming a mature adult.

The whole project takes decades, but satisfaction increases as you pass every signpost and turning point along the way. There is a sharp divide between seniors who have aged to be regretful, unfulfilled, and depressed and seniors who look back with contentment and inner satisfaction. By age seventy the die is cast. But the maturing process starts with a vision of the goal. To us, the goal is embodied in the phrase *core self.* This is the part of you that shapes your reality, placing you at the center of experiences you personally create.

HOW IT FEELS TO HAVE A CORE SELF

You know that you are real.

You don't feel controlled by others.

You don't live for approval; you aren't crushed by disapproval.

You have long-range goals to work toward.

You work through difficult situations for your own sense of dignity and self-worth.

You give respect and receive it from others.

You understand your own emotional life. You are not swayed by other people's emotions.

You feel safe in the world and like where you belong.

Life has brought a certain wisdom.

To have a core self is to be the author of your own story; it is the exact opposite of being a victim, who must live a life authored by others. Because it sets goals, your core self walks ahead of you. You cannot expect to capture it today, any more than a kindergarten child can capture being a college freshman. The reason we use *core self* instead of simply saying *mature self* is that maturity has a bad name; it tends to connote someone whose life is boring and staid. In truth, your life journey becomes far more exciting if you are following a vision that inspires you year after year. Visions create the opportunity for fulfillment; therefore the core self is the source of enormous power, from which your future grows.

Breaking out: To begin, shift your allegiance away from superficial activity and toward the deep project of becoming a completely authentic, mature person. Sit down and write out your personal vision. Aim for the highest goals you can imagine that would bring fulfillment. Seek out people who share the same vision and are achieving success. Once you know where you are headed, the path unfolds with its own inner guidance. Allow this to happen; your unfolding potential needs reinforcement day by day.

4. Align yourself with the flow of evolution, or personal growth.

This chapter on the evolving brain has made the point that future evolution is a choice. Your brain is not bound by Darwinian evolution. Your survival isn't at stake; your fulfillment is. Choosing to grow automatically means that you are facing into the unknown. Guidance on the path is wobbly at first. Everyone contains some kind of insecurity that gradually gives way to elements of self-possession and true knowledge.

But without evolution there would be no path, only aimless wandering. Evolution is a cosmic force. It's the reason that drifting clouds of stardust led to life on Earth. It's the source of all creativity and intelligence. Every good idea you've ever had, every moment of insight or *aha!* proves that evolution is invisibly at work, guiding life from behind the scenes.

We strongly believe that the universe supports everyone's evolution, but at the same time you can guide your own growth. Desire is the key. We all desire more and better things for ourselves. If those more and better things are good for your growth, then you are guiding your own evolution. If what you desire is likely to help others, the odds are higher it will be attained.

WHAT MAKES A DESIRE EVOLUTIONARY?

It doesn't repeat the past but feels fresh and new.

It helps more people than just you.

It brings a glow of contentment.

It fulfills a deep wish.

You don't regret it.

It opens easily and naturally.

You aren't fighting with yourself or with outside forces.

Fulfilling it will serve others as well as yourself.

It opens a greater field of action.

It expands your awareness as fulfillment grows.

Desire is an untrustworthy guide if all you think about is what feels good at the moment and what feels bad. You need a larger frame of reference. Indian culture makes a distinction between *Dharma* and *Adharma*. Dharma includes whatever naturally upholds life: happiness, truth, duty, virtue, wonder, worship, reverence, appreciation, nonviolence, love, self-respect. For the individual, the flow of evolution supports all these qualities, but you must choose them first.

On the other hand, there are bad choices, *Adharma*, that do not support life naturally: anger, violence, fear, control, dogmatism, harsh skepticism, unvirtuous acts, self-indulgence, the conditioning of habit, prejudice, addiction, intolerance, and unconsciousness in general. What unites the world's wisdom traditions, East and West, is knowing what is dharmic and what is adharmic. One leads to enlightenment and freedom; the other leads to greater suffering and bondage.

Breaking out: Follow the dharmic path. Dharma is the ultimate power, because if evolution is supporting all of creation, it easily supports you, a single individual. Honestly look at your everyday life and the choices you are making. Ask yourself how to increase the dharmic choices and decrease the adharmic ones. Step by step, follow up your conviction to evolve.

5. Trust in a higher power that transcends everyday reality.

Nothing that we've described so far will come true without a higher vision of reality. For the moment, let's leave aside religion and any reference to God. Far more important is the chance to go beyond a passive role in order to embrace the role of reality maker. Whatever is holding you in a powerless state, if you are destined to be stuck there, you aren't going to regain power.

Fortunately, the power to go beyond suffering has always existed; it is your birthright. To have even a speck of consciousness is to be connected to the infinite consciousness that supports evolution, creativity, and intelligence. None of these things is accidental or a privilege handed out to the lucky few. When you ask to be connected to a higher reality, the connection is made.

GLIMPSES OF HIGHER REALITY

You feel watched over and protected.

You feel cared for.

You recognize blessings in your life that feel like acts of grace.

You feel gratitude for being alive.

Nature fills you with wonder and awe.

You have had some experience of seeing or sensing a subtle light.

A divine presence has touched you personally.

You've experienced moments of pure ecstasy.

Miracles seem possible.

You sense a higher purpose in your life. Nothing has been accidental.

How close is higher reality? To use a metaphor, imagine that you are caught in a net. All nets have holes, so find one, and jump through it. Higher reality will be there waiting for you.

One wife of a domineering husband found herself feeling stifled and powerless. She had never worked outside the home, devoting twenty years to raising a family. But she jumped out of the net when she discovered painting. It was far more than a pastime. Art was an escape route, and as she found buyers who appreciated her paintings, a shift occurred inside. Her reality picture went from *I am trapped and can do nothing* to *I must be worth more than I imagined, because look at this beautiful thing I created.*

Breaking out: Escape routes exist everywhere in consciousness. All you need to do is be aware of the potentials hidden in your awareness and latch on to them. What are the possibilities in life that you hungered to fulfill but never did? These are the choices you need to revisit. If you pursue something you deeply cherish, higher reality will reconnect with you. This new connection registers "in here" as joy and curiosity, a whetted appetite for tomorrow. It registers "out there" as ever-expanding possibilities that support you when you least expect it.

Everything we've discussed is a kind of escape route in the end. All escape routes lead back to the core self, the person who was born to be a reality maker. That person is unconcerned with individual power. What really counts extends far beyond the individual: it is the glory of creation, the beauty of nature, the heart qualities of love and compassion, the mental power to discover new things, and the unexpected epiphanies that bring the presence of God—these universal aspects are your true source of power. They are you, and you are all of them.

WHERE
HAPPINESS LIVES

I f you can make reality, what would an ideal reality look like? To begin
with, it would look personal. As your brain constantly remodels itself,
it conforms to what you, as a unique individual, want from life. Happiness? You might suppose that this would top the list. But it turns out
that the desire for happiness immediately exposes a serious weakness.
Although we are designed to be reality makers, most people aren't especially skilled at making their personal reality a happy one.

Only recently, with the rise of a new specialty known as positive psychology, has happiness been closely studied. The findings
are mixed. When people are asked to predict what will make them
happy, they list things that seem obvious: money, marriage, and
children. But the particulars do not bear this out. Taking care of
small children is actually a source of high stress for young mothers.
Half of marriages end in divorce. Money buys happiness only up to
the point where it makes the material things in life secure. Poverty is
certainly a source of unhappiness, but so is money, since once people
have enough of it to secure the basic necessities, adding extra money
doesn't make them happier—in fact, the added responsibility, along
with fear of losing their money, often does the opposite.

The overall picture, surprisingly, is that even when people get what they wanted, most of them aren't as happy as they thought they would be. Climbing to the top of your profession, winning a sports title, or making a million dollars looks great as a future goal, but those who arrive at such goals report that the dream was better than the achievement. Competition can turn into a never-ending process, and its rewards decrease over time. (A study of top tennis champions found that they were motivated less by the joy of winning than by the fear and disappointment of losing.) What about people who fantasize about striking it rich and never having to work for the rest of their lives? A study of lottery winners, for whom this fantasy came true, found that the majority said winning actually made their lives worse. Some couldn't handle the money and lost it; others found their relationships strained or descended into reckless behavior like gambling and making fly-by-night investments. All were plagued by strangers and relatives asking incessantly for handouts.

If people are such bad predictors of how to be happy, what can we do?

The current fashion in psychology holds that happiness can never be permanent. Polls have found that around 80 percent of Americans—and often more—report that they are happy. But when examined individually, researchers find that each person experiences only flashes of happiness, temporary states of well-being that are not permanent at all. Therefore many psychologists contend that we stumble on happiness without knowing how to achieve it.

But we diverge from this view. Our feeling is that the problem lies with reality making. If you have more skill at creating your own personal reality, then permanent happiness will follow.

MOVING TOWARD LASTING HAPPINESS

DO

Give of yourself. Take care of others, and care *for* them.
Work at something you love.

Set worthy long-range goals that will take years
  to achieve.

Be open-minded.

Have emotional resilience.

Learn from the past, and then put it behind you. Live for
  the present.

Plan for the future without anxiety, fear, or dread.

Develop close, warm social bonds.

DON'T

Hitch your happiness to external rewards.

Postpone being happy until sometime in the future.

Expect someone else to make you happy.

Equate happiness with momentary pleasure.

Pursue more and more stimulation.

Allow your emotions to become habitual and stuck.

Close yourself off from new experiences.

Ignore the signals of inner tension and conflict.

Dwell on the past or live in fear of the future.

In a consumer-driven society, it's all too easy to fall into doing all the don'ts on this list, because they share the same element: linking happiness with temporary pleasure and external rewards. But here's the story of a man named Brendon Grimshaw, who must have a very finely honed instinct for happiness, since he created his own personal paradise.

Paradise Is Personal

Grimshaw, born in Devonshire, England, was working as a journalist in South Africa when he walked away from his job in 1973. He had taken the extraordinary step of buying his own tropical island—Moyenne Island, in the Seychelles chain located between India and Africa—for 8,000 pounds, or around $12,000. He owned

Moyenne for nine years, then made the decisive leap to live there, all alone with a native Seychelles helper. What faced this modern Robinson Crusoe was daunting. He did the opposite of lounging on the beach. The undergrowth was so dense on the island when he arrived that falling coconuts didn't hit the ground.

Grimshaw set about clearing the underbrush, and as he did, he let the island speak to him—that's his description of how he approached new plantings. He found that mahogany trees thrived on Moyenne, so he imported a few at first, and he now has seven hundred, reaching sixty to seventy feet high. But they are a fraction of the sixteen thousand trees he has planted by hand. He gave refuge to the rare giant tortoise of the Seychelles and has 120 of them. Birds flock to this protected sanctuary, and two thousand of them are new to the island.

In 2007 Grimshaw's helper died, so at eighty-six he is the sole caretaker of his island, for which he has reputedly been offered $50 million, which he turned down. He shakes his head when visitors see the mahogany trees solely as a source of wood for furniture and the pristine beaches as a haven for rich vacationers, who have been visiting the Seychelles more and more. Moyenne will remain a preserve after his death. In person, Grimshaw looks sunburned and weathered as he tramps around in his bush hat and shorts, but he is remarkably alive, too. His state of contentment can be traced back, almost item for item, to the things on our list. He gave of himself while working at something he loved. He set a goal that took years to achieve. He was dependent on nothing and no one outside himself to give him constant approval.

About the only aspect of lasting happiness that is absent from this story is social bonding. But for some people, solitude is richer company than society, as it is for Grimshaw. His life also conforms to the concept of a fully integrated brain, one that merges every need that the brain is designed to serve. These include:

Connecting with the natural world
Being useful
Exercising your body
Finding work that satisfies
Fulfilling your life purpose
Aiming beyond your limited ego-self

No separate region of the brain oversees the merging of these needs to make a fully developed person. It takes the entire brain, acting as an integrated whole. Happiness is then rooted in the feeling that you are complete. The most credible version of the fully integrated brain is the one laid out by a Harvard-trained psychiatrist, Dr. Daniel J. Siegel, now at UCLA, who has made a career of examining the neurobiology of human moods and mental states. Siegel has pioneered the fascinating study of how our subjective states correlate with the brain. What sets him apart from researchers who perform thousands of brain scans to discover how the brain lights up during certain states is that Siegel's aim is therapeutic. He wants his patients to get better. The route to healing, he maintains, is to trace symptoms like depression, obsession, anxiety, and so on back to the exact brain region that is causing a block.

Since every thought and feeling must register in the brain, it makes sense that psychological symptoms like depression and anxiety are indications of faulty wiring—that is, a neural pathway has been laid down that continues to repeat the undesirable symptoms or behavior. It functions like a microchip that has no choice but to reiterate the same signal. But neural "wiring" can be changed, such as through therapy—Siegel uses talk therapy in conjunction with his brain-centered theory.

Siegel's goal is a healthy brain that sustains a person's well-being. As he sees it, the brain needs healthy nutrition every day. His approach is in accord with ours, since he prescribes a "healthy mind platter" of daily nourishment, with the idea that a healthy mind leads to a healthy brain. On his mind platter, Siegel and colleague David Rock place seven "dishes."

Sleep time
Physical time
Focus time
Time in
Down time
Play time
Connecting time

Years of brain research lie behind these simple prescriptions, but as science learns more and more that all aspects of life lead back to the brain, the nutrition offered on Siegel's mind platter could be far more important to the body than any conventional advice. Your brain has an enormous talent for integration, but more than that, if it is used holistically, the brain thrives on putting everything together.

Doing the Work

Let's consider the benefits of these seven nutrients, which we will divide into *inner work* and *outer work*.

Inner work: Sleep time, focus time, time in, down time

Inner work is the area of subjective experience. A healthy day, as viewed by the brain, follows a natural cycle. You have had enough sleep to be adequately rested. You focus intensely, with enough down time to let the brain rebalance and find an easy resting place. You have down time for doing no mental work—letting the mind and brain simply be. And you set aside a period for what many Western-ers neglect: going inward through meditation or self-reflection. This is the most precious time, actually, since it opens the way for evolution and growth.

What goes on in your inner world? Most people, if they are being honest, devote eight hours at work to focused activity. Then they go home, find a way to relax, and distract themselves until it is bedtime. If work is unsatisfying, they focus only as much as they have to, and their real pleasure "in here" comes from pure distraction, diverting their frustrations with television, video games, tobacco, and alcohol.

But as Siegel points out, the brain is caught between two dysfunctional states: chaos and rigidity. If your inner world is chaotic, you feel confused. Conflicting emotions are hard to resolve; impulses are hard to resist. If the chaos gets out of hand, fear and hostility can roam your mind at will, and sometimes you aren't responsible for your own behavior. We casually describe chaotic people with ill-defined terms like *flighty, "a mess," hysterical, out of control, spacey*—all attempts to get at a state of disordered confusion.

Rigidity counters chaos in the wrong way. Rigid people have clamped down. Their behavior is set along fixed patterns. They deny themselves any spontaneity, and they resent (while secretly fearing) anyone who is spontaneously happy. Rigidity leads to ritualistic behavior—like long-married couples who repeat the same arguments year after year. Taken to an extreme, rigidity leads to severe judgments against others, enforcing rules with harsh punishments. We casually refer to rigid people as anal, uptight, hard asses, tight asses, "fascists," morality police—these terms have in common a constricted, tightly organized approach to life. But considered without judgment, the suffering that results from a tightly wound inner world is real. Rigidity, because it feels safer than being chaotic, can gain social approval. Every society has a law-and-order party; none has a *carpe diem* (live for the moment) party.

Siegel places the integrated brain in the middle between chaos and rigidity; it is the true solution to both, which is why inner work is necessary. We will be more specific about the spiritual side of inner work later. The key thing to absorb here is the natural cycle that every day should follow. For example, sleep research has indicated that all but the tiniest fraction of adults needs eight to nine hours of good sleep every night. After a good night's sleep, the brain needs to wake up on its own, taking the time it needs to switch from the chemical state of sleep to the chemical state of being awake, which is entirely different.

It's a myth that sleep can be shortchanged. From the brain's viewpoint, getting six hours of sleep during the week is a permanent loss. You can't fully make it up by sleeping in on the weekend. Waking up with an alarm clock is also detrimental. The brain transitions from deep sleep in a series of waves, each one getting closer to full wakefulness. You rise up to light sleep, then back down again, several times during this process, and as you do, your brain secretes a bit more of the chemicals needed to be awake. If you cut the process short, you may tell yourself that you are awake, but in fact you aren't. Schoolkids who stay up late playing video games will sit through homeroom and first period the next day basically in their sleep. Adults who have slept six hours can function reasonably well for the first four to six hours of their workday, but after that there is a steep falling off. The loss of one hour's sleep impairs driving skills—about as much as taking two alcoholic drinks.

Most people are aware of the importance of sleep, but as a society we don't do what is good for us in this area. We are chronically sleep deprived and even proud of the fact, since it indicates a life on the go and total dedication to our work. But the mind platter indicates that true dedication would consist of balancing the brain for optimal performance, which means taking seriously time in, down time, and sleep time. Our overworked, overstimulated society ignores these three areas.

Outer work: Physical time, play time, connecting time

This is the area of outward activity. Inner and outer work cannot be strictly divided from each other, since all brain processes are inner and all behavior is outer. Speaking generally, however, when you interact with someone else, you are doing outer work. You chat, gossip, and bond. You go to restaurants and conduct a hopeful crawl through bars. You build a family and find things to do together. As many sociologists have pointed out, this area of life used to dominate everyday existence, at a time when families sat around the fire of an evening and ate every meal together.

That's no longer true. Families today are often loose constellations. Contact is intermittent and rushed. Everyone has their own space. Activity is scattered all around town, not confined to the home. Cars have made everyone mobile, but central heating may be the most powerful force in shaping modern society. In the past, bedrooms were cold chambers where you retreated for sleep. Otherwise, you spent the evening in the one or two rooms of the house that had a fire going. The kitchen, now considered the heart of a home, was the province of servants in all but the poorest households.

Physical separation makes outer work harder. We are seeing new brain changes in the digital generation, who have adapted to physical separation more than ever. By spending hours focused on video games and social networking, young people are expanding one set of skills—the eye-hand coordination needed for video games and the technical expertise for computers—while neglecting the neural pathways for interacting with people face to face. It is telling that being on Facebook, which is essentially a constantly updated photo album with commentary, is considered a "relationship." Actual personal contact isn't necessary.

But if you leave judgment out of the picture, social networking represents a new kind of shared mind, a global brain with activity

that connects hundreds of millions of people. The sense of connection that comes from instantly tweeting your thoughts is real, and
the sense of belonging to something bigger than yourself is real, too,
as when news of the turbulent events of the Arab Spring in 2011
went around the world in real time. There is much optimism about
social networks changing the world for the better. In repressive soci-
eties in the Middle East, some feel that the future is a race between
the mullahs and the iPad—in other words, a contest between traditional repressive forces and the technology that frees people's minds.

If connecting time is exploding in the digital era and play time
can be had with the touch of a Wii box, the neglected ingredient is
usually physical time. The brain needs physical activity, even though
we think of this organ, naturally, as mental. But because it monitors
and controls the body, your brain participates in physical stimulation. The things that decrease physical time are all around us, and
unfortunately they are all detrimental to the brain. Being depressed
keeps people shut inside and inactive. Replacing outdoor exercise
with compulsive computer activity puts the body in a sedentary
state, which is unhealthy. Being totally sedentary increases the risk
of almost every lifestyle disease, including heart attack and stroke.

The message to get out and exercise has increasingly fallen on
deaf ears—guilty deaf ears—as Americans and Europeans grow
more sedentary and gain weight. According to a 2011 report from
the Centers for Disease Control, a quarter of U.S. adults report that
they devote no time to physical activity. The number increases to 30
percent across the South and Appalachia—for them, "couch potato"
has become a dismal reality—while only 20 percent meet the recommended amount of physical activity. For reference, federal guidelines recommend that adults between eighteen and sixty-four get a
total of two and a half hours of moderate activity or one hour and
fifteen minutes of intensive activity per week. The recommendation
goes up for children and adolescents (ages six to seventeen), who
should be doing at least an hour of intensive activity a day, which

would typically be provided in physical education class at school. Participation in PE has been on a steady decline, however.

Residents are most likely to be physically active in parts of the Northeast, the West Coast, Colorado, and Minnesota. (One reason for regional variation may be the influence of peers. If someone in your peer group goes for a run, you are more likely to do the same.) But because the data were self-reported, people may have overstated their level of physical activity, which means that these statistics are overly optimistic.

One result is almost predestined. A third of adult Americans are overweight compared to where they should be, and another third have become obese. Exercise has a direct brain connection, when you consider what it actually does. The benefits to increased cardiovascular health are well known, and obviously exercise gives you better muscle tone. What we tend to overlook are the feedback loops that connect the brain to every cell in the body. Therefore when you throw a ball, run on a treadmill, or jog along the shore, billions of cells are "seeing" the outside world. The chemicals transmitted from the brain are acting the way sense organs do, making contact with the outside world and offering stimulation from that world.

This is why the jump from being sedentary to doing a minimal amount of exercise—such as walking, light gardening, and climbing the stairs instead of taking the elevator—is so healthy. (Each step of doing more exercise adds benefit to your health, but the single largest benefit comes from getting off the couch in the first place.) Your cells want to be part of the world. Such a statement would have sounded far-fetched in the past. Mainstream physicians back then regarded the mind-body connection as suspect. As a result, medicine adopted a hostile attitude toward "soft" psychological explanations, considering drugs and surgery all-important. Drugs and surgery require a simple cause-and-effect relationship between disease X and cause Y. The cold virus causes colds, and pneumococcus bacteria cause tuberculosis. The breakdown of simple cause-and-effect is vital for us, however. It leads to the idea of the fully integrated brain as essential to health—a super brain.

Let's look more closely at the path that mind-body integration had to travel when it came to a disorder that afflicts society en masse: heart disease.

Making the Link

The link to the brain was slow in coming. In the 1950s America began to experience an alarming increase in premature heart attacks, those that occur primarily in men between the age of forty and sixty. As deaths from heart disease and stroke skyrocketed, physicians began to see more and more men complaining of chest pain, which turned out too often to be angina pectoris, a primary symptom of blocked coronary arteries. At the turn of the century the renowned William Osler, one of the founders of Johns Hopkins Medical School, was on record noting that a doctor in general practice would hardly see one case of angina a month. Suddenly it became common to see half a dozen a day.

Scrambling for an explanation for the epidemic, cardiologists focused on a physical cause, as the drastic increase of fats in the American diet compared with that of our grandparents, who ate far more whole grains and vegetables. One factor that seemed eminently scientific: cholesterol. A massive public campaign was launched to get the public to consume a diet lower in red meats, eggs, and other sources of cholesterol. The campaign may not have been a huge success, since the national diet remains high in fat, but cholesterol became a scary word (overlooking that your body produces 80 percent of the cholesterol in the bloodstream and that this steroid is absolutely necessary for building cell membranes); a billion-dollar industry has grown up around reducing the "bad" blood fats and increasing the "good" ones. From the beginning, nobody seriously considered the brain as a cause of heart attacks. It was left out of the loop because no model existed for how the brain could transmit messages to heart cells, and the term *stress* was barely being mentioned.

As it happened, some experts were dubious about cholesterol from the start; they pointed out that soldiers who were casualties in the Korean War had been autopsied, and it was found that even in their early twenties, their coronary arteries already contained enough plaque to lead to a heart attack. Why didn't heart attacks arrive until much later? No one knew. It was suggested, when analyzing the extensive data provided by the Framingham Heart Study, that men in their twenties who faced their childhood psychological issues were better protected against premature heart attacks than those who didn't. But this was no time for such "soft" explanations.

No one believed that you could think your way to a heart attack. The decision was made to back cholesterol as a ready-made villain. (We won't go into the problems facing the cholesterol hypothesis, except to mention that the cholesterol you ingest doesn't necessarily lead to high blood cholesterol—the physiological picture is complex and growing more so every decade.) The brain wasn't taken into account even when a psychological argument finally became popular, the argument over Type A and Type B personalities. Type A people were tense, demanding, perfectionists, prone to anger and impatience, and addicted to control. As a result, the theory went, Type As were more prone to heart attacks than Type Bs, who were relaxed, tolerant, even-tempered, patient, and more accepting of mistakes. Type As did seem much more likely to create stress. (There was a quip at the time about having a Type A boss: he's not the kind who gets a heart attack—he's the kind who gives them.) As it turned out, actually identifying and testing who was Type A or Type B proved elusive; now instead of "personality," medicine speaks of Type A or Type B behavior.

Once stress and behavior entered the picture, you'd think that the brain must have become a major player, but it didn't. There was still no model for explaining how an external stress could enter the body and find a physical pathway to the cells.

By the late 1970s such a pathway began to emerge with the discovery of *messenger molecules*, a class of chemicals that turn moods, stress, and disorders like depression into something physical. The public began to hear about brain cells in detail as biologists named the neuropeptides and neurotransmitters that leap across the synapses, the gaps between neurons. *Serotonin* and *dopamine* became household terms, with links to chemical imbalances in the brain (for example, too much serotonin or too little dopamine). A great era of discovery was at hand, and the decisive step came when it was found that these chemicals not only leap across the synapse but course through the bloodstream. Every cell in the body contains receptors that are like keyholes, and the brain's chemical messengers are the keys that precisely fit the hole. To simplify a complex model, the brain was telling the entire body about its thoughts, sensations, moods, and general health. The link between psyche and soma, mind and body, had been made at last.

It's now generally accepted that psychological factors contribute to the risk of developing heart disease. The list of factors includes

Depression
Anxiety
Personality traits
Type A behavior
Hostility
Social isolation
Chronic stress
Acute stress

Your heart participates in mental distress and can react with clogged arteries—an amazing finding compared with what was medically acceptable several decades ago. Instead of focusing merely on disease prevention, health experts began to speak of something more positive, far-reaching, and holistic: well-being. The brain became the centerpiece of a chemical symphony orchestra with hundreds of billions of cells joining in, and when they were in total harmony, the result was increased well-being; meanwhile chemical disharmony led to higher risk for disease, early aging, depression, and decreased immune function, as well as all the lifestyle disorders—the list keeps growing beyond heart attacks and strokes to include obesity, Type 2 diabetes, and probably many if not most cancers.

We want to follow the implications of this new trend as far as they will take us. We fully endorse Siegel's concept of a healthy mind leading to a healthy brain. A mind that reaches for higher consciousness brings even more benefits, especially in terms of happiness. When you use the guidelines for inner and outer work, you are providing your brain with the right nutrients.

Yet happiness will still be elusive. Nutrients do not create meaning. They don't define a vision or set a long-range goal. Those are your tasks as a reality maker. You have another frontier to cross before you reach the most desirable thing of all, a personal paradise that no one can ever take away from you.

SUPER BRAIN SOLUTIONS

SELF-HEALING

By now, as was not the case two decades ago, the mind-body connection has been proved over and over. It is an established fact, and yet moving to the next step—using the mind to heal the body—remains elusive and controversial. No single practice ensures results—we have no mind-body equivalent to a magic bullet. Even though spontaneous remissions have been observed in almost every form of cancer, and even though some of the deadliest malignancies like melanoma have the highest rates of spontaneous cures, the phenomenon is rare (estimated by some surveys at fewer than twenty-five cases per year in the United States, although there is widespread doubt over any such measurement).

Self-healing has nothing to do with seeking a miracle cure or trying to be that one patient in ten thousand who recovers, to their doctor's amazement. Healing is as natural as breathing, and therefore the key to healing is a lifestyle that optimizes what the body is already doing.

A HEALING LIFESTYLE

Practice the recommended amount of moderate healthy exercise.

Keep your weight down.

Reduce your stress.

Attend to psychological issues like depression and anxiety.

Get adequate sleep.

Don't worry about vitamin and mineral supplements if your diet is healthily balanced (unless you have a condition like anemia or osteoporosis, where a doctor may advise a specific supplement).

Avoid toxic substances like alcohol and nicotine.

Reduce animal fats in your diet.

Strengthen the mind-body connection.

These guidelines all sound familiar, but that doesn't reduce their effectiveness. The best healing is prevention; there is no getting around it. But the last item on the list—strengthening the mind-body connection—may be the most powerful, and for most people it's new territory. We've covered the mind platter of daily activity that benefits the brain. Here we'd like to enter the more elusive matter of healing through the mind-body connection.

Being Your Own Placebo

The most studied technique of mind-body healing has been the placebo effect. *Placebo* is Latin for "I shall please." It's a good way of describing how the placebo effect works. A doctor offers a patient a powerful drug, with the assurance that it will relieve the patient's symptoms, and the patient, as promised, gets relief. But in reality the doctor has prescribed a harmless, inert sugar pill. (The effect isn't limited to drugs, which is important to remember: anything you believe in can act as a placebo.) Where did the patient's relief come from? It came from the mind telling the body to get well. To do that, the mind must first be convinced that healing is about to occur.

The big problem with the placebo effect, which is known to operate in 30 percent of cases on average, is that the first step is deception. The doctor misleads the patient, which has proven to be a serious ethical roadblock. No ethical physician would regularly deny the best care to a patient, offering instead innocuous substitutes, even though in some cases (such as mild to moderate depression) studies show that drugs are likely to be no more effective than a placebo. This means, by the way, that many drugs share the unpredictability of the placebo effect. The notion that pharmaceuticals act the same way for all patients is a myth. The placebo effect, contrary to widespread suspicion, is a "real" cure. Pain is diminished; symptoms are alleviated.

Now let's ask the most important question: Can you be your own placebo without using deception? If you give yourself a sugar pill, you know in advance that it offers no relief. Is that the end of it? By no means. Self-healing through the placebo effect depends upon freeing your mind from doubts—without deceiving yourself. People need to know more about the mind-body connection, not less.

Being your own placebo is the same as freeing up the healing system through messages from the brain. All healing is, in the end, self-healing. Physicians aid the body's intricate healing system (which coordinates immune cells, inflammation, hormones, genes, and much else), but actual healing takes place in an unknown way.

When it comes to the mind-body connection, healing should involve the following basic conditions:

The mind is contributing to getting well.
The mind doesn't contribute to getting sick.
The body is in constant communication with the mind.
This communication benefits both the physical and mental
    aspects of being well.
Once the person receives treatment that he trusts, he lets go
    and allows the healing response to proceed naturally.

When the placebo effect works, all five aspects are involved. The patient's mind cooperates with the treatment and trusts it. The body is aware of this trust. There is open communication, and as a result, cells throughout the body participate in a healing response. The healing system as a whole is incredibly complex and all but impossible to explain as a whole. We only know how parts of it operate, such as antibodies and the immune response to infection.

How can we bring about these five conditions consciously? At the very least, we shouldn't be fighting them with fear, doubt, skepticism, hopelessness, and despair. Those states convey their own chemical messages to the body. When you believe that a sugar pill is going to cure you, those healing messages will begin to have an effect. But we cannot say that the 30 percent who benefit from the placebo effect are doing something right while the remaining 70 percent aren't. Everyone's medical history is different; the healing system remains too murky to be accurately measured. Deep negative feelings, if they are blocking the placebo effect (by no means a certainty), are complex and frequently unconscious, so the difference here isn't simple.

The greatest promise lies in the fact that a mental intention of "I shall please" is known to work. Being your own placebo requires applying the same conditions as in a classic placebo response:

1. You trust what is happening.
2. You deal with doubt and fear.
3. You don't send conflicting messages that get tangled with each other.
4. You have opened the channels of mind-body communication.
5. You let go of your intention and allow the healing system to do its work.

When a symptom is minor, such as a cut finger or a bruise, everyone finds it easy to let go and stop interfering. The mind doesn't intrude with doubts and fears. But in serious illness, doubts and fears play a marked role, which is why a practice like meditation or going to group counseling has been shown to help. Sharing your anxiety with others in the same predicament is one way to begin to clear it.

It's also helpful to follow your healthiest instincts. Many of us deal with illness through misleading processes like wishful thinking and denial. Our fears lead us into blind alleys of false hope. In such cases, the mind isn't really alert to what the body is saying, and vice versa. The atmosphere is clouded. To trust what your body is telling you requires experience. You need a certain amount of mind-body training, and that takes time. It's well documented, for example, that a positive lifestyle, which includes exercise, diet, and meditation, reduces heart disease. The combination allows the body to reduce the plaque that blocks coronary arteries. But the improvement doesn't happen overnight. It requires patience, diligence, and time.

This is the opposite of receiving a diagnosis of cancer, panicking, and running desperately after any possible cure. Becoming an instant convert to prayer or meditation under the duress of disease is almost always futile. Fear becomes worse when you are seriously ill, yet dealing with anxiety is far more effective if you attend to it years before you ever get sick. The mind-body connection has to be strengthened before trouble arrives.

The very important task of becoming aware of your body doesn't have to be boring. You mostly need mind and body to make friends again, to go back to their natural alliance. One way to do so is to sit quietly with your eyes closed and simply feel the body.

Let any sensation come to the surface. Don't react to the sensation, whether pleasant or unpleasant. Just relax and be aware of it. Notice where the sensation is coming from. You won't have only one sensation or feeling. You will find that your awareness goes from place to place, one moment noticing your foot or your stomach, your chest or your neck.

This simple exercise is a mind-body reconnection. Too many people are in the habit of paying attention to only the most gross signals from their bodies, such as acute pain, stiffness, nausea, and other hard-to-ignore discomforts. What you want to do is to increase your sensitivity and your trust at the same time. Your body knows at a subtle level where dis-ease and discomfort are. It sends signals at every moment, and such signals are not to be feared.

Even if you consciously ignore what is happening in your cells, just below your awareness unconscious information is being exchanged. When the federal government recently decided that annual mammograms are not necessary for younger women, one consideration was that 22 percent of small breast tumors resolved themselves, disappearing spontaneously. So an automatic reaction of fear, even in the face of possible cancer, is unrealistic at the level of the healing system. Your immune system eliminates thousands of abnormal cells every day. Everyone has tumor-suppressing genes, although how they can be triggered is as yet unknown.

The future of self-healing will unfold from the proven fact that every cell in the body knows, through chemical messengers, what every other cell is doing. Bringing your conscious mind into the loop adds to this communication. Advanced yogis can alter their involuntary responses at will, such as lowering their heart rate and breathing to very low levels, or increasing skin temperature in a very precise way. You and I have the same abilities, although we don't consciously use them. You can follow an exercise to make a spot on the palm of your hand grow warmer, and it will happen, even though you never used that ability before.

We can venture that the placebo effect falls into the same category. It's a voluntary response that we could use if only we learned how to. The healing system seems to be involuntary. You don't have to think in order to heal a cut or a bruise. But the fact that some patients can make their own pain go away when given a sugar pill implies, very strongly, that intention makes a difference in healing. We aren't talking about positive thinking, which is often too superficial and masks underlying negativity. Instead, we are encouraging a lifestyle that bonds a deeper mind-body connection.

Note: The brain's connection to the placebo effect is crucial but has only recently been studied in depth. Because a book is a public discussion read by all kinds of people with all kinds of health issues, let us be clear. We are not advising anyone to stop conventional medical treatment or to reject medical help. The placebo effect remains mysterious, and this section is simply exploring that mystery, not giving you a how-to for miracle self-cures.

PART 3

MYSTERY AND PROMISE

THE
ANTI-AGING BRAIN

To unlock any new promise that super brain holds, we must first solve an old mystery. No mystery is older—or greater—than aging. Until very recently, only magical elixirs, potions, or the fountain of youth were possible escapes from the ravages of age. Resorting to magic shows how baffled the mind was. Growing old is universal, with reprieve for no one, and yet medically speaking, no one ever dies of old age. Death occurs when at least one key system of the body breaks down, and then the rest of the body goes with it. The respiratory system is almost always involved; the immediate cause of death for most of us will be that we have stopped breathing. But a person can just as effectively die of heart or kidney failure. Meanwhile, virtually all of the genetic material in the body may be viable at the moment when the key system fails.

How do we prevent that one critical system from bringing down everything else? You would have to pay attention to the whole body for a lifetime. Prediction is extremely difficult. Several factors prevent anyone from seeing in advance where the aging process will ultimately lead.

Uncertainty 1: Aging is very slow.

It begins around age thirty and progresses at roughly 1 percent a year. This slowness prevents us from actually observing a cell as it ages. We see the effects only after years have passed. Nor are these effects uniform. For every aspect of physical and mental deterioration, some people actually get better with age. By getting enough exercise, they may become stronger than they were when they were young. For a small, fortunate few, at age ninety, memory can improve rather than decline. Aging is like a ragged army, in which some cells advance ahead of others, but the whole army moves at a snail's pace and with great stealth.

Uncertainty 2: Aging is unique.

Everyone ages differently. Identical twins who are born with the same DNA will have completely different genetic profiles at age seventy. Their chromosomes won't have changed, but decades of life experience will have caused the activity of their genes to be switched on and off in a unique pattern. The regulation of each cell, minute by minute for thousands of days, makes their bodies age in unpredictable ways. In general, we are genetic duplicates of one another at the moment of birth but entirely one of a kind at the moment of death.

Uncertainty 3: Aging is invisible.

The aspects of growing old that you see in the mirror—gray hair, wrinkles, sagging skin, and so on—indicate that something is going on at the cellular level. But cells are immensely complicated, undergoing thousands of chemical reactions per second. These reactions are fixed and automatic. Bonding occurs among various molecules, dependent on the atomic properties of the elements that make up the body, principally the big six—carbon, hydrogen, nitrogen, oxygen, phosphorous, and sulfur. If these atoms are shaken up in a beaker, they will perform automatic reactions in a few thousandths of a second. On its own, phosphorous is so volatile that in a fiery collision with oxygen, it will explode. But over billions of years, living organisms developed incredibly intricate combinations that prevent such crude interactions. The phosphorous in your cells isn't explosive. It enters into an organic chemical known as ATP, adenosine triphosphate, a key component in binding enzymes and transferring energy.

A biologist could spend a lifetime studying how just this one complex molecule operates inside a cell, yet the controller of each reaction remains unseen and unknown. As long as a cell is functioning smoothly, no one needs to see the controller. A kind of chemical intelligence is clearly at work, and it's enough to say that DNA, because it contains the code of life, is the beginning and the end of everything that goes on inside a cell. But thanks to aging, cells stop functioning with complete efficiency, and then the invisible element raises its head. Atoms do not have the capacity to go wrong, but cells do. Why and how is not predictable—it is traceable only after a wrong turn has been taken.

All these uncertainties lead to a single conclusion. There is no alternative to paying attention to your whole body for your lifetime. But this is the very thing that people find almost impossible to do. Our lives are full of contrasts, and we are addicted to its ups and downs. Walking the straight and narrow sounds boring. It implies a kind of stifling Puritanism, where self-denial is the rule and pleasure the exception. The real challenge, as we see it, is to make lifetime well-being so desirable that it stops being a penance.

How to begin? No matter which approach you take to anti-aging, your brain is involved. No cell in the body is an island—all are receiving an unbroken stream of messages from the central nervous system. Certain messages are good for cells, and others are bad. Eating a cheeseburger every day sends one kind of message; eating steamed broccoli sends another. Being happily married sends a different message from being lonely and isolated. Clearly you want to send messages to tell every cell not to age. Therein lies the promise.

If you can maximize the positive messages and minimize the negative ones, anti-aging becomes a real possibility.

It turns out that anti-aging is a gigantic feedback loop that lasts a lifetime. The term *feedback loop* keeps returning in this book because science is discovering more and more about how these loops work. In 2010 an exciting joint study from the University of California at Davis and UC San Francisco revealed that meditation leads to an increase in a crucial enzyme called telomerase. At the end of every chromosome is a repetitive chemical structure called a telomere, which acts like the period at the end of a sentence—it closes off the chromosome's DNA and helps to keep it intact. In recent years the fraying of telomeres has been connected to the breakdown of the body as it ages. Due to imperfect cell division, telomeres get shorter, and the risk emerges that stress will degrade a cell's genetic code. Having healthy telomeres seems to be important, and therefore it's good news that meditation can increase the enzyme that replenishes telomeres, telomerase.

This research sounds highly technical, mainly of interest to cell biologists. But the UC study went a step further and showed that the psychological benefits of meditation are linked to telomerase. High telomerase levels, which also seem to be supported by exercise and a healthy diet, are part of a feedback loop that results, surprisingly enough, in a sense of personal well-being and the ability to cope with stress. This one finding helps to cement the most basic tenet of mind-body medicine: that every cell is eavesdropping on the brain. A kidney cell doesn't think in words; it doesn't say to itself, *I've had a horrible day at work. The stress is killing me.* But it is participating wordlessly in that thought. Meditation brings a sense of well-being to the mind, while silently spreading the same feeling, via a chemical like telomerase, to your DNA. Nothing is excluded from the feedback loop.

The mind-body connection is real, and choices make a difference. With those two facts in place, the anti-aging brain holds untold promise.

Prevention and Risks

Without knowing why we age, medicine has taken the approach that aging is like a disease. Germs cause cellular damage, and so does growing old. It's sensible to focus on keeping your body healthy and functioning. The physical side of anti-aging is similar to prevention programs for any lifestyle disorder. Let's review the main points. They will seem familiar after decades of public health campaigns—yet they are still a vital part of your physical well-being.

HOW TO REDUCE THE RISKS OF AGING

Eat a balanced diet, cutting back fats, sugar, and processed foods. The preferred diet is Mediterranean: olive oil instead of butter, fish (or soy-based sources of protein) instead of red meat, whole grains, legumes, mixed nuts, fresh fruits, and whole vegetables to provide plenty of fiber.

Avoid overeating.

Exercise moderately for at least one hour three times a week.

Don't smoke.

Drink alcohol, preferably red wine, in moderation, if at all.

Wear a seat belt.

Take steps to prevent household accidents (from slippery floors, steep stairs, fire hazards, icy sidewalks, etc.).

Get a good night's sleep. It may also be helpful as you grow older to take an afternoon nap.

Keep regular habits.

In terms of prevention, the physical side of anti-aging keeps being refined. Take the issue of obesity, which has now reached epidemic proportions in America and Western Europe. Being overweight has long been recognized as a risk factor for many disorders, including heart disease, hypertension, and Type 2 diabetes. But now a specific kind of fat, belly fat, is being targeted as the most damaging kind. Fat isn't inert like the fat in a stick of butter. It is constantly active, and belly fat sends out hormonal signals that are damaging to the body, as well as altering metabolic balance. Unfortunately, exercise alone will not get rid of belly fat. A general weight-loss and exercise program is needed; eating sufficient fiber also seems to help combat belly fat.

Given our wealth of refined knowledge, the real problem lies elsewhere, with compliance. Knowing what's good for you and doing it are two different things. Exercise is a constant drumbeat in prevention advice, yet we are becoming an increasingly sedentary society. Fewer than 20 percent of adults get the amount of exercise recommended for good health; one out of every ten meals is eaten at McDonald's, where the food is high in fat and sugar and almost absent fiber and whole vegetables.

Compliance is difficult when your brain is wired to make the wrong choices. Certain tastes, for example—especially salty, sweet, and sour—are so immediately attractive that we gravitate toward them. With repetition, these tastes become the ones we prefer. Given enough repetition, they become the tastes we reach for automatically, victimized by unconscious habit. (The snack-food industry has a term—*munch rhythm*—to describe the automatic way a person keeps putting popcorn, potato chips, or peanuts into his mouth without stopping until the bag is empty. This is the ultimate unconscious behavior, considered highly desirable among snack-food purveyors but disastrous for anyone's diet.)

It's futile for health experts to nag the public year after year to change its ways and then expect compliance. It's still less effective for you to nag yourself. The worse you feel about yourself, the more likely you are to drift into discouragement. Once you feel discouraged, two things happen. First, you grow numb, bored with fighting yourself. Second, you seek to palliate your discomfort, usually through distractions. You watch television or seek out quick fixes of enjoyment by eating salty snacks and sweets. In this way, the effort to do better ends up doing worse. If nagging actually worked, we'd be a nation of joggers elbowing each other to get at the organic produce section of the supermarket.

Aging is a very long process. A class in stress management, a few months of yoga, going vegetarian for a while—these are blips on the screen when it comes to aging's slow creep. Clearly, to prevent aging, we have to crack the problem of noncompliance.

Conscious Lifestyle Choices

The secret to compliance isn't exerting more willpower or beating yourself up for not being perfect. The secret is *changing without force.* Anything you force yourself to do will eventually fail. Anti-aging isn't built in a day. Whatever you do now, you must keep doing for decades. So let's stop thinking in terms of discipline and self-control. Some people are prevention saints—they consume only one tablespoon of total fat per day in their diet, because that's the ideal amount for heart health. They ignore wind and rain and get in five hours of vigorous exercise a week. Saints are inspiring to the rest of us, but deep down they are also discouraging, because they remind us that we are a hundred miles from being saintly ourselves.

Change without force is certainly possible. To achieve it, you need to create a matrix for making better choices. By matrix, we simply mean your setup for daily living. Everyone has a matrix already. Some people's matrices make positive choices much easier than do others. A cupboard that contains no snack foods would be part of such a matrix. A house without a television or video games would be another, but if you are jogging every day because you have no entertainment at home, you aren't being good to yourself. In the end, the physical side is secondary. A matrix is more substantial and sustainable. That's why we surround ourselves with support for the behavior we like best.

The real secret is to live inside a matrix where the mind feels free to choose the right thing instead of feeling compelled to choose the wrong thing.

MATRIX FOR A POSITIVE LIFESTYLE

Have good friends.

Don't isolate yourself.

Sustain a lifelong companionship with a spouse or partner.

Engage socially in worthwhile projects.

Be close with people who have a good lifestyle—habits are contagious.

Follow a purpose in life.

Leave time for play and relaxation.

Keep up satisfying sexual activity.

Address issues around anger.

Practice stress management.

Deal with the reactive mind's harmful effects: When you have a negative reaction, stop, stand back, take a few deep breaths, and observe how you're feeling.

We've already covered many of these items in our discussion of the ideal lifestyle for your brain, but the same ones have also been correlated with longevity. One thing that links them is very basic: success comes when people act together; failure tends to happen alone. Having a spouse or life partner who keeps an eye on your diet ("Haven't you already eaten a cookie today? Have a carrot") is better than wandering the supermarket aisles alone and impulsively grabbing a week's worth of frozen dinners. A friend who goes to the gym with you three times a week gives you more incentive than all the promises you make to yourself as you watch *Sunday Night Football*. It's important to establish your matrix early and keep it going. Studies have shown that losing a spouse suddenly leads to isolation, depression, higher risk for disease, and shortened life span. But if you have a social network beyond your spouse, you have a cushion against these baleful consequences.

The most crippling aspects of aging tend to involve inertia. That is, we keep doing what we've always done. Starting in late middle age, new things gradually fall by the wayside. Passivity overtakes us; we lose our motivation. Countless old people find themselves stranded by inertia.

Deepak recalls a couple who foundered when the wife turned fifty. She looked upon that birthday as a milestone, a new starting point. With her children ready for college and her job secure, she wanted to open new areas of life that she hadn't been able to explore while family duties were pulling her away from some of her deepest dreams.

"My husband and I had a yearly ritual," she says. "We took a long weekend alone and evaluated our marriage. It was quite systematic. We made a list of each element in our relationship, including sex, work, hidden agendas, and resentments. We are both very organized people, and just before I turned fifty, we rated each aspect of our marriage and discovered that we scored at least eight out of ten in each category. I felt happy and secure."

So it came as a shock when this woman sat down one night and revealed her plans for making the marriage succeed for the next twenty years. Her husband, who was highly successful in business, turned to her and said, "I don't want to change. Why bother? We'll get old. I see us sitting in easy chairs waiting for the kids to call."

Unseen by her, her husband had been succumbing to creeping inertia. His whole life centered on his work first and foremost; when retirement came, as he saw it, there was nothing left to achieve. "I've already done whatever I'm going to do. Why try and repeat the past? It's hard enough to keep doing the same thing over and over."

This couple went into counseling, but their views were too divergent. On the eve of divorce, both were disappointed yet quite content in their own choice. The wife felt free to build a new life based on new aspirations. The husband was content to rest on his laurels and look nostalgically back on the past. Each was an intelligent person with high self-esteem and a sense of confidence.

But with the passage of time, as fifty turns to sixty and then to seventy and eighty, which of them has made a better choice? The wife is building on the matrix that sustained her for her first five decades; the husband is trusting that time will take care of itself. There are no guarantees in life, but most psychologists would predict that she has a better chance for longevity and, more than that, a better chance to feel fulfilled as she ages.

Linking with Immortality

So far, we've covered the key aspects of the "new old age," the rubric applied to a social movement that advocates positive aging. For two decades the image of old age has been shifting dramatically. No one expects anymore to be put on the shelf at sixty-five. A large proportion of baby boomers don't see retirement in their future. Growing old is being pushed further off than ever. In a sense, this is the positive side effect of living in such a youth-oriented culture. No one wants to face being no longer young. The latest wave of seniors are making positive lifestyle changes, if not quickly enough (and with not enough equality. The increase in longevity that has benefited the top half of income earners in America hasn't extended to the bottom half, where life expectancy remains closer to seventy than to eighty, where the upper half are quickly heading).

What is the next step, then? We feel that anti-aging needs to look beyond the physical and even beyond the psychological. The best life is rooted in a vision of fulfillment, so that it's the life one would want to extend. It's hard to have a vision that defies aging, because for untold generations human beings have looked around, and what do they see? They see that all creatures grow old and die. But this universal observation is not in fact true. In a very real sense cells are immortal, or at least as close to immortal as living organisms can be. Can this be the clue to a new and higher vision of life?

The original blue-green algae that evolved billions of years ago are still with us. They never die but simply divide and keep dividing. This is also true of single-celled organisms like amoebae and paramecia, found in pond water. Adverse circumstances certainly kill off primitive life forms by the billions, but accidents of nature aren't the same as natural life span. The natural life span of many cells is unlimited. Only when they gather into complex plants and animals do cells face the prospect of death. A red blood corpuscle that dies at three months, a white blood cell that dies as soon as it consumes an invading germ, a skin cell blowing away in the wind—all are living out a natural life span. But the body integrates hundreds of different life expectancies—as many life spans as there are tissue types. Even then there is tremendous leeway and flexibility. Stem cells exist even in the oldest living human, with the potential to mature into fresh new cells.

The cells in your body have retained all the mechanisms of primitive life-forms, including cell division, but they have also kept evolving. Complex creatures like mammals have added life-saving inventions that primitive organisms don't possess, such as an immune system. A human body faces many threats that do not trouble blue-green algae, yet each one has been met, over the course of evolution, with highly creative ways to defend, cope, and survive. The human mind took over from cellular evolution a long time ago. The single greatest benefit to longevity, for example, may have been sanitation—sewage treatment and clean water were evolutionary leaps for mankind (and the loss of them around the world, as sewage and water supplies become contaminated, poses a grave danger to hundreds of millions of people). Medicine, of course, is a major, ongoing way to extend life.

Each of us is caught between two forces that contend for our personal future, the force of evolution, which extends life longer and longer, and the force of entropy, which causes physical things to decay over time. Aging is a highly complicated form of entropy; it's not as simple as a star using up its fuel, collapsing upon itself, and exploding in one last dramatic death throe as a nova or supernova.

The situation is so complex, in fact, that each person can choose which side, creation or destruction, to favor. Entropy isn't destiny. There is no reason why you can't choose to favor evolution every day. Ultimately, our true link with immortality is through evolution, which has driven creation for 13.8 billion years since the Big Bang. On some early spring day when the trees have found the courage to believe that winter is over, go outside and pluck a stem of new growth from a flowering tree or sprouting rose bush. Examine it, and you will notice that every tender shoot has a growing tip that reaches into the unknown. As vulnerable as that tender shoot may look, it is repeating the act of creation that has endured forever. This is the physical sign of life's faith in itself.

In a very real sense, you are the growing tip of the universe. An eternity of time, longer than the life of the oldest galaxy, has conspired to arrive at this moment in one person's existence. Where is the universe heading in the next moment? Only you can choose. You are responsible for your own growth, and yet the choice is more than personal. The timeless has delivered itself into our hands. It awaits your decision, and wherever you go next, reality will follow. If you think we are exaggerating—or even being outlandish—consider what your cells are doing. Without their link to immortality, life can't exist.

SUPER BRAIN SOLUTIONS

MAXIMUM LONGEVITY

Whenever a cell ages, you age. This is the biological bottom line. Yet cells are powerfully designed, over the course of evolution, to survive. They are linked to chemical processes that are literally immortal, or at least as old as the universe. Ironically, even if you do everything wrong in terms of lifestyle—chronically smoking, stuffing your body with fats and sugar, never exercising—the same brain that is implicated in your horrible choices is itself trying to stay immortal. Brain cells are like all other cells, waging a successful campaign to defeat time, and this campaign is taking place second by second, from the moment of conception in the womb.

We've been waxing a bit philosophical, yet there are specific ways that you can live a vision of maximum longevity. Winning the genetic lottery is rare. Various research projects have looked at specific mutations that allow some family lines of Ashkenazi Jews to live beyond one hundred— fathers, mothers, brothers, and sisters all become centenarians. (Previously there had been no historical documentation of a family where more than one person in a generation lived to be a hundred.) The key seems to be that their genes make them immune to plaque buildup in their arteries, the chief cause of heart attack and stroke. At this point, however, the prospect of transferring this genetic advantage to others is remote.

In the general population, life span keeps extending in developed countries. Japanese women are the most long-lived on earth. The rising life expectancy that Americans have enjoyed every decade is well understood—improved sanitation and medical care have been critical. Childhood infectious diseases have been curbed, and most recently major improvements in emergency care for heart attacks and recovery programs for stroke have been important. The decline in smoking has also benefited life expectancy. The last two hurdles are probably lack of exercise and obesity. In other words, as long as people take prevention seriously and make positive lifestyle changes, the physical basis for living a long time is taken care of. Only the rarest individuals will become centenarians (about 1 in 30,000), while more and more of us will live in good health into our eighties and nineties.

The standard view is that in order to make a significant advance over the present situation, we need to find a cure for cancer and Alzheimer's. Certainly both are scourges of old age. Heart disease remains the leading cause of death in the United States—despite advances in treatment, medicine has still not discovered what causes it. The plaque collected in the coronary arteries may look like debris clogging a drain. However, it takes microscopic wounds or lesions in the smooth lining of a blood vessel to give tiny particles of fatty deposits a place to lodge. This process begins when we are still quite young, and although the risk factors are well known, such as high cholesterol, smoking, a sedentary lifestyle, type A behaviors, and high stress, risks are not the same as causes.

At present, longevity presents a confusing picture among genes, risk factors, and drugs, the latter being favored by the pharmaceutical companies. Older people on average take seven prescribed medications, all of which have side effects. Pills are easy to pop—and easy for the doctor to prescribe—but over the last decade, leading drug treatments for depression, heart disease, and arthritis have come under scrutiny for being less effective, or more dangerous, than they were promoted to be. If anything, the focus on drugs has lessened the public incentive to practice prevention, which has no side effects and proven benefits.

We'd like to discuss the most personalized approach to longevity, which is tuning in to your body. This requires self-awareness. On one hand, you have a lifetime of likes and dislikes, habits, beliefs, and conditioning. On the other hand, you have the wisdom that has evolved in every cell. Anti-aging is a matter of making these two halves mesh. This is a perfect example of survival of the wisest.

THE WISDOM OF CELLS
SEVEN LESSONS IN LONGEVITY

1. Cells share and cooperate. No cell lives in isolation.
2. Cells are self-healing.
3. The life of the cell demands constant nutrition.
4. Cells are always dynamic—they die if they get stuck.
5. The balance between inner and outer worlds is always maintained.
6. Toxins and disease organisms are immediately spotted and defended against.
7. Death is an accepted part of the cell's life cycle.

Cells became wise over billions of years of evolution; you can become just as wise by using the gifts of self-awareness, paying attention to how biology has solved some of the deepest issues that you face in everyday life.

1. Cells share and cooperate. No cell lives in isolation.

You are part of the human community, and coexistence is the most natural and healthy way to live. Cells don't struggle with this truism. They benefited enormously by coming together to form tissues and organs—your brain is the most spectacular evidence of that. But we're all tempted to strike out on our own, driven by ego to amass more and more for ourselves, adding close family but excluding almost everyone else. (One memorable book on how to get rich looked at the lives of self-made millionaires and came to a depressing conclusion: most were "stingy S.O.B.'s.") Cells are not so misguided that they look out for number one.

We aren't trying to deliver a moral lesson here. Some fascinating research has shown that social connections are mysteriously contagious. Sifting through the massive data bank of the Framingham Heart Study, which has examined risk factors related to heart attacks for thirty-two years, social scientists made a startling discovery. Obesity, one of the major risks for heart disease, spreads like a virus. In the social network of family, coworkers, and friends, simply relating to someone with a weight problem makes it more likely that you will have one. "According to the data, if one person became obese, the likelihood that his friend would follow suit increased by 57 percent. (This means that the network is far more predictive of obesity than the presence of genes associated with the condition.) If a sibling became obese, the chance that another sibling would become obese increased by 40 percent, while an obese spouse increased the likelihood that the other spouse would become obese by 37 percent."

Using statistical methods that linked 12,067 residents of Framingham, Massachusetts, the researchers found that the viruslike behavior of obesity could also be applied to other risks like smoking or depression. If you have a friend who smokes, the likelihood that you will smoke increases, while having a friend who quits smoking increases your likelihood of making the same positive change. But the most enigmatic aspect is that you don't have to relate to someone else directly. If your friend has a friend, unknown to you, who is obese or depressed or a smoker, your chances of acquiring those habits increases, if by a tiny margin.

Other social scientists have found these correlations unacceptable, but so far no one has found a better model for explaining how behavior gets passed along. The point is that placing yourself in a positive social context is good both physically and mentally. In a way not fully understood, our cells understand what it means to do good.

A classic Harvard psychology study from the 1980s asked subjects to watch a film depicting Mother Teresa's work with sick and orphaned children in Calcutta. As they watched, their blood pressure and heart rate went down.

Going a step further, in 2008 a study by University of Michigan social psychologist Sara Konrath examined the longevity of 10,000 state residents who had participated in a health study going back to their graduation from high school in 1957. Konrath focused on those who had done volunteer work in the past ten years, and her findings are fascinating. Individuals who volunteered lived longer than nonvolunteers. Of the 2,384 nonvolunteers, 4.3 percent died between 2004 and 2008, but only 1.6 percent of the altruistic volunteers had died.

The key word is *altruistic*. People were asked why they volunteered, and not every answer involved altruism. Some of the participants' motives were more oriented toward others, such as "I feel it is important to help others" or "Volunteering is an important activity to the people I know best." Other respondents, however, had more self-oriented reasons for volunteering, such as "Volunteering is a good escape from my own troubles," or "Volunteering makes me feel better about myself." Those people who said they volunteered for their own personal satisfaction had nearly the same mortality rate (4 percent) as people who did not volunteer at all. This is just one example among many to support that invisible traits in the mind-body system have physical consequences. Your cells know who you are and what motivates you. The Michigan research was the first to show that what motivates volunteers can have an impact on life expectancy.

To move from ego-centered selfishness to social sharing is a process. The steps look something like this:

I want to be liked and accepted.
If I keep everything to myself, others reject me.

We can succeed together or fail separately.
I can afford to share. It doesn't hurt. It feels good, in fact.
When I give, I find that I receive.
The more I have, the more I can afford to give.
Strangely, there's a fullness in giving away more and more.
The most fulfilling kind of giving is to give of myself.
I find the deepest connection comes from generosity of spirit.

As with everything in life, the path from the first step to the last isn't a straight line but a zigzag that is different for each person. A three-year-old learning to share toys can't comprehend what generosity of spirit is. Some people never comprehend it, no matter how many years pass. Yet building a self follows this arc, matching a cell's natural design, which involves sharing and cooperation as a matter of survival. At the level of the self, survival isn't usually the issue. The issue is the rewards you receive through bonding and connecting, the basic process that makes for a peaceful society.

2. Cells are self-healing.

When you are self-aware, you learn how to repair your own damage. This comes naturally to cells, although healing is still one of the most complex and baffling bodily processes. We only know that it exists and that life depends upon it. Cells are fortunate not to have to think about healing. They spot any damage, and instantly the repair mechanism sets to work. At the mind-body level, there is a basic parallel. When we say that time heals all wounds, we are talking about an automatic process, however painful it may be. Grief runs its course, for example, without anyone knowing how shattered emotions are actually healed.

But much healing isn't automatic, as we know from those who never recover from grief. Most of the time healing is a conscious activity. You look inside and ask "How am I doing?" all the time. There is no guarantee that you will find the key, and when your inner damage is sore and aching, even looking at it can be too much. Self-healing means overcoming the pain and finding a way to become whole again. The path looks something like this:

> *I hurt. Somebody help me.*
> *I hurt again. Somebody help me again.*
> *Why won't the hurt go away? If I don't look at it, maybe it will go away.*
> *I tried to distract myself, but I really need to get at this pain inside.*
> *I can stand to look at what's wrong.*
> *Maybe there's something I can do for myself.*
> *This pain could be telling me something—but what?*
> *I think I understand, and now the pain is beginning to subside.*
> *I feel incredibly relieved. Healing is possible.*
> *I trust my ability to heal.*

A small child crying out for Mommy to come and help has no other resources. A child can't comprehend the final stage, *I trust my ability to heal.* But healing is part of the vast feedback loop that holds mind and body together. The more you experience even a moment of trying to heal yourself, the greater your ability grows. The triumph over one's deepest wounds is a spiritual triumph. Without it, life would be cruel, since wounds are unavoidable. Only by building a self can you prove to yourself that life isn't cruel so long as victory over pain is possible. Through self-awareness, you realize that healing is one of the most powerful forces sustaining your life.

3. The life of the cell demands constant nutrition.

Cells survive by having complete trust that the universe will support them. So solid is this trust that a typical cell stores no more than three or four seconds of reserve food and oxygen. Nourishment is always coming in. Using that certainty, a cell can devote all its time and energy to the things that cause life to move forward: growing, reproducing, healing, and running its own internal machinery. At the same time, cells don't pick and choose what is good for them. All nourishment is good. There's no time to make mistakes or to flirt with risky lifestyles.

Here is a piece of wisdom honored more in the breach than in the observance. In our culture excitement, risk, and danger are positive words, while balance, proportion, and moderation feel impossibly dull. We take it as our birthright to experiment with rebellion. So we have every temptation to ignore the benefits of a balanced life, and while we experiment, our cells suffer. But wisdom has more than one teaching. Everyone cherishes the right to make mistakes, and evolution is very forgiving. You can always retrace your steps and lead a life that is more nourishing. The important thing is to know what is most nourishing to you personally and put your energies there.

When you do that, passion becomes part of balance. Presumably cells are passionate about life already—they do all they can to thrive and multiply, after all. So nourish yourself with the three things that would increase your passion for life. It's worthwhile to sit down and actually write the three things down, tucking the list in your wallet to take out and remind yourself whenever you need to. Leaving specifics aside, your nourishment needs to embrace mind and body. Therefore, your list should include:

1. Your highest vision.
2. Your deepest love.
3. Your longest reach.

Vision gives you purpose and meaning. Love gives you vibrant emotions and lasting passion. A long reach gives you a challenge that will take years to meet. Taken all together, these three primal elements lead to true happiness. As with every aspect of wisdom, you have a path to follow when it comes to nourishing your life. It looks something like this.

> *I guess I'm happy enough. My life is as good as that of the person sitting next to me.*
> *I only wish my day wasn't so routine and predictable.*
> *Just beneath the surface I have secret dreams.*
> *Maybe I don't have to be afraid to stretch myself.*
> *I deserve more quality and happiness in my life.*
> *I will risk following my bliss.*
> *My aspirations are starting to come true.*
> *It's unbelievable, but the universe is on my side.*

This is an arc of growing trust, the kind of trust that comes naturally to cells but that gets compromised in our own lives. For most people, trust runs into a roadblock early on. They lose the simple trust of a child, who depends on his parents to feed, clothe, and support his existence. A transition comes when a new sort of trust—self-reliance—enters. During this transition, a person learns to stop making trust external (*I trust Mommy and Daddy*) and begins to trust internally (*I trust myself*). Clearly this difficult transition involves many setbacks. So it takes constant awareness to keep evolving. The only true nourishment that lasts a lifetime comes from within. If you keep placing your trust in other people, they can be taken away from you. But if you trust yourself, there is no such threat. The path leads from *I can do this myself* to *I am enough* and finally *I am supported by the universe*. No path is more rewarding or more sublime.

4. Cells are always dynamic—they die if they get stuck.

Cells are immune to many troubles that plague everyday life—they have to be in order to survive—and one blessing is that they never get stuck. A cell's whole world is the bloodstream, which is a super-highway of chemicals brimming with traffic. Blood looks uniform to the naked eye, a slightly viscous, warm, crimson liquid. But at the molecular level, it teems with change. A cell never knows exactly what the superhighway is going to bring next. The blood chemistry of a soldier in battle, a patient just diagnosed with cancer, a yogi sitting in a Himalayan cave, and a newborn baby is entirely unique.

Cells, in response to an ever-changing world, adapt instantly. The brain is forced to be the most adaptable, since all operations in the body, however minuscule, are reported back to it. Therefore, if you get stuck in a behavior, habit, or belief that refuses to budge, you are hampering your brain. It took a long time for medical science to accept how serious stuckness can be. Twenty years ago some early mind-body studies looked for correlations between psychology and disease. Many doctors suspected, without scientific proof, that some patients had personalities that made them more susceptible to cancer in particular. Results did emerge—a so-called "disease personality" was marked by emotional repression and a general uptightness. But there was no "cancer personality." Therefore it wasn't of much use to discover that your psychology might put you at a vague, general risk of almost any disease from the common cold to rheumatoid arthritis and heart attacks.

But we can make use of this finding by turning it on its head. Instead of trying to pinpoint the kind of behavior that makes cancer more likely, we can focus on not getting stuck, since we know that brain cells—and all the body's other cells—are designed to be dynamic, flexible, and constantly alert to change. Learning that change is your friend doesn't come naturally to everyone, and as we age, resistance to change grows. The path to follow typically looks something like this.

I am what I am. Nobody has the right to change me.
Familiarity creates my comfort zone.

My daily routine is beginning to seem stale.

I see people who do more things than I do. Maybe I've suppressed my curiosity.

I can't expect life to bring new things to me. I need to motivate myself.

New things are beginning to be enjoyable.

It's possible to create a comfort zone in the midst of change.

I love my dynamic life—it keeps me feeling vital.

Cells don't have to follow this path; evolution has ensured that dynamism is simply a fact of life. It's on the personal level that you must confront your stuckness. In the end, the reason is natural and basic: you are designed to evolve because that's how your body operates. Your cooperation with nature may meet with resistance at first, but if you press on, it's the easiest way to live and thrive.

5. The balance between inner and outer worlds is always maintained.

Cells don't get hung up about their inner world. They aren't neurotic or anxious about the future. They harbor no regrets (although they certainly do carry the scars of the past—ask the liver of an alcoholic or the stomach lining of a chronic worrier). Because they don't complain, it's easy to assume that cells have no inner life, but they do. The dividing line between inner and outer is the cell's outer membrane. In many ways it is the cell's miniature brain, because the cell receives all its messages at the receptor sites that crowd the cell membrane by the thousands. These receptors let some messages in and keep others out. Like floating lily pads, they open out to the world but have roots that reach beneath the surface.

On the inside, these roots allow various messages to go where they are needed. If you experience denial or repression, or the censorship of certain feelings and the eruption of others, or if you feel the tug of addiction and the inflexibility of habits, all those things can be traced to the cell membrane. Receptors are constantly changing, fulfilling the need to keep the inner and outer world in balance. This is yet another aspect of the gift of adaptability. Deepak likes to say that we don't just have experiences; we metabolize them. Every experience turns into a coded chemical signal that will alter the life of your cells, either in a small way or in a big way, either for a few minutes or for years at a time.

Trouble arises when a person seals off his inner world and fails to match it to the outer world. There are two extremes here. At one extreme are psychotics, whose only reality is their distorted thinking and hallucinations. At the other extreme are sociopaths, who have no conscience and barely any inner world; their whole focus is exploiting other people "out there." In between these two poles lies a vast array of behaviors. The inner and outer worlds become imbalanced through all kinds of defensive mechanisms. In other words, we insert a kind of screen that separates the outer world from the way we react to it. The kinds of screens that people put up include the following:

Denial—refusing to face how you really feel when things
    go wrong.
Repression—growing numb to feelings so that events "out
    there" can't hurt you.
Inhibition—clamping down on feelings, using the logic
    that diminished feelings are safer as well as more
    acceptable to society.
Mania—letting feelings run wild, without regard for the
    repercussions that will come your way from society; the
    opposite of inhibition.
Victimization—denying yourself pleasure because others
    won't give it to you, or accepting the burden of pain
    because you feel you deserve to.

Control—seeking to place fences around both the inner
   and outer world so that neither can break through your
   boundaries.
Domination—using force to keep others in an inferior
   position while you indulge in your own power fantasy.

What would it be like for you to live without these screens?
In a word, you would have emotional resilience. Studies of people
who have managed to live in good health to the age of one hundred indicate that their biggest secret is the ability to remain
resilient. Centenarians have suffered the same setbacks and dis-
appointments as everyone else, but they seem to bounce back
more easily and to let the burden of the past weigh lightly on
them. Emotional resilience implies that defensive mechanisms
are not very present, because when they are, a person holds on
to old hurts, harbors secret resentments, and incorporates stress
rather than throwing it off. Your body pays the price for every
defense you put up.

Cells don't act in any of these distorted ways. Instead, there is
in-flow and out-flow, the natural rhythm of life. The cell's inner
response matches external events. To restore this rhythm in yourself
requires awareness. Everyone has psychological baggage, and our
tendency is either to protect our inner self from more hurt or to ignore our inner life because it's too messy to face. The path that leads
to a balance between "in here" and "out there" might go something
like the following:

This feels bad. I don't want to deal with it.
It's not safe to show how I feel.
The world is a scary place. Everyone has a right to protect
   themselves.
I'll deal with my issues tomorrow.
Things don't seem to be getting better on their own.

*Maybe I need to face my hidden attitudes and suppressed
feelings.
I've looked inside, and there's a lot of work to do. But it isn't as
scary as I imagined.
It's a relief to let go of old issues.
I'm beginning to feel more comfortable in the world and a lot
safer.*

6. Toxins and disease organisms are immediately spotted and defended against.

If cells had an opinion about the way we conduct our lives, they
would no doubt express amazement at how much we tolerate toxins.
By nature, cells instantly expel toxic substances or counteract them.
The immune system's chief assignment is to separate harmful invad-
ers from harmless ones. The kidneys' assignment is to filter toxins
from the blood. A vast array of bacterial flora are in your intestines
and need to be there (taking an antibiotic will indiscriminately wipe
out most of the bacteria in your body and will throw off your digestion for a while, perhaps dramatically), and an equally vast range of
biochemicals course through the blood. The immune system and
kidneys have evolved to tell the good from the bad. Your body's
intelligence is finely attuned to toxicity and guards against it. The
same lesson has been harder for human beings to learn.

When mainstream medicine ignored the campaign for a more
natural diet and against food additives, it did a disservice to the
public welfare. Ever since the meat and dairy industries began mas-
sively adding hormones to speed up the production of meat and
increase drastically the amount of milk that dairy cows give, suspi-
cious changes have occurred in public health, such as the early onset
of menstruation in young girls and a rise in breast cancer. (Breast
tissue is highly sensitive to foreign substances and can easily mistake
them for hormonal signals.) Even today the average physician has
minimal education in nutrition and diet. But doctors should have

joined the campaign against potentially toxic contamination of our air, water, and food.

Populations with contaminated water and inadequate sewage are prone to all kinds of epidemics and shortened life span. But we have not yet studied the correlation of life expectancy with additives in the "normal" American diet. The government monitors the use of pesticides and insecticides by law, yet it rarely pursues and prosecutes violations. Huge market forces at work promote fast food, quick-to-market beef, high sugar content, and a wide range of preservatives. But no one needs to wait for studies to pinpoint which additive is toxic and which isn't. A high-fat and sugar-laden diet is already risky. Caution is the best attitude; eating a natural diet makes the best sense. Why not favor the least toxicity in your diet that you can reasonably achieve?

This shouldn't turn into a rationale for extremes. To date, no study has shown that people who obsessively take large amounts of supplements or who rigorously eat organic food live longer than people who eat a normal balanced diet. *Toxin* is a scary word, but a balanced approach is better than total purity motivated by fear. Pesticides and insecticides are legally mandated to degrade by the time food comes to market, and they are washed off when produce is processed for sale—in any event, washing your fruits and vegetables at home should be standard practice. It's sensible not to entirely trust the food industry, who reassure us that we don't ingest enough preservatives, additives, and pesticides to harm us. Over a lifetime, you are what you eat. That is warning enough.

The campaign for a better diet is part of the overall trend toward higher compliance (if only it would move faster), and the bigger problems are the invisible toxins that degrade well-being. They, too, are well publicized: stress, anxiety, depression, domestic violence, and physical and emotional abuse. You can't see or taste these toxins, but the same difficulty—noncompliance—is at work. People put up with toxic lifestyles far too much. They behave in ways that impact their bodies very badly, or they endure similar behavior from family, friends, and coworkers. The solution is awareness, honestly looking in the mirror and finding a way to expel invisible toxins from your life. The pathway looks something like this:

I'm strong and healthy. I can eat anything I want.
Nothing seems to be going wrong.
"Natural" is for ex-hippies and people who worry too much.
I've looked into it, and there are more toxins than I thought.
Better to be safe than sorry.
I have to change today if I want to be healthy tomorrow.
I can wean myself off processed foods if I try.
I deserve a state of well-being. It will take effort, but it's worth it.

Ridding your life of invisible toxins follows a different pathway but not that dissimilar. You move from thinking *I can put up with this* to *My life is being damaged* and finally to *I deserve a state of well-being.* Rationales and inertia are powerful things. We can spend years putting up with toxins because our minds find reasons not to change. Recognize how powerful these forces are, and respect them. You don't need to mount a frontal assault in an attempt to purify your life. It's good enough to evolve in the right direction. The wisdom that took billions of years to evolve in cells deserves a few years of serious consideration from you.

7. Death is an accepted part of the cell's life cycle.
Cells manage something we can only envy and barely understand: they put all their energy into remaining alive, and yet they are not afraid to die. We've already touched on apoptosis, which is programmed death that genetically tells a cell when it is time to die. But most of the time cells divide rather than die in the way we fear our own death—they defy mortality when they turn themselves into a new generation of cells. Reincarnation happens before your very

eyes, if you are watching cell mitosis through a microscope. Human beings have a more unsettled attitude toward dying, but over the past few decades—accelerating after Elisabeth Kübler-Ross's ground-breaking 1969 book, *On Death and Dying*—our social attitudes have become less fearful.

The wisdom of cells correlates perfectly with the world's great teachers of wisdom. Death is not the equal and opposite of life. It is part of life, which overarches everything. Whatever is born must die, and yet in the cosmic scheme, to die is only a transition to another kind of life. Renewal is nature's constant theme. These themes are controversial when people compare their religious beliefs and when they war over dogmatic truth. But cells are not theological, and neither is Nature as a whole.

A skeptic will fire back at any faith-based view of life, contending that the universe is cold and impersonal, ruled by random events, and ultimately indifferent to human existence. Strangely, the contest between faith and skepticism doesn't seem to impact how a person approaches their own mortality. Coming to terms with dying is so personal that it transcends belief. There are devout believers who quake with fear at the prospect of dying, and skeptics who face it with equanimity. The essential point, first made on a wide scale by Kübler-Ross, is that dying is a process that passes through stages. By now those stages are familiar: grief, denial, anger, bargaining, depression, and acceptance. (Deepak knows two sisters who attended their eighty-nine-year-old mother as she was going through hospice care. Each sister sat on one side of the bed, taking turns reading aloud from *On Death and Dying*, hoping to offer solace to their mother, who was listening quietly with eyes closed. Suddenly it dawned on them that she had died. Spontaneously one sister exclaimed, "But you're only on stage four!")

In the intervening years, disagreement has arisen over whether Kübler-Ross correctly described the stages of dying and over the order in which they occur. But the bigger lesson is that dying should be as dynamic as living, an experience that evolves as you enter it for yourself. Some cultures, such as Tibetan Buddhism, offer extensive preparation for death and a highly detailed theology of various heavens and hells (although these *bardos* are better thought of as states of consciousness after the physical body has been left behind). The West has no such tradition—except among Native Americans—and each person must consider the issue of death on his or her own. But consider it we must. If you are afraid of death, it is bad for your body, not because death looms so darkly but because fear of any kind is toxic.

The picture of feedback loops constantly sending messages to your cells is inescapable. The good news is that the sting of death is largely psychological, and so you can remove it. Nature is on your side. The vast majority of dying patients have come to terms with it; hospice workers often note that it is the family of the dying person who have the greatest anxiety and suffer from the most stress. Moreover it's too casual, and mistaken, to link aging and death. Aging happens to the body; death happens to the self. Therefore, the person who has the strongest sense of self, who has deeply investigated the great question of "Who am I?" is likely to have the most calmness in the face of death.

We will have much more to say about how to reach your true or core self. It's a vital issue since the world's wisdom traditions declare that death cannot touch the true self—and that truth is what Saint Paul means by "dying unto death." Here we want to emphasize that dying is a natural part of life, as every cell in our body already experiences. The path to making peace with death might look something like the following:

I don't think about death. It's pointless.
The main thing is to live your life right this minute.
Anyway, I secretly don't believe I will grow old and die.
To be honest, I don't think about dying because it's too scary.

I've seen death, of a friend, family member, or pet. I know I
  have to face it someday.
I am beginning to feel calmer about the whole issue. I can look at
  death without running away.
Dying happens to everyone. It's better to approach it calmly,
  with eyes open.
I've felt the first serious twinges of mortality. It's time to face it.
I find I am actually interested in what death is all about.
It's possible to embrace dying as a natural stage of life—
  and I have.

Achieving wisdom is a lifetime project. We are encouraged by
the "new old age" and by studies that indicate the positive side of
aging, which can be grouped under the rubric of maturity. Older
people tend to underperform on tests of memory and IQ compared
with younger people, but in areas of lifetime experience, they outperform them. This is especially so in tests that ask you to make a
decision about a challenging situation like firing an employee, telling a friend that his spouse is cheating, or facing the diagnosis of
a serious illness in the family. What is needed in such situations is
maturity, and although emotional intelligence comes into play, no
single aspect of IQ equates with maturity. You must live a life to acquire it. Why not live that life in accord with evolution, as embodied
in your cells?

THE
ENLIGHTENED BRAIN

What would it be like to be enlightened? Is the soul within reach? Can God be experienced personally? For many, answering these questions is like capturing a unicorn, a beautiful dream that never turns into reality. The unicorn represented perfect grace in the Middle Ages. The pure white horse with a braided horn coming out of its forehead was a symbol of Christ, and capturing it was an inner journey to find God. Myth can become reality if you find the right path.

Enlightenment also involves taking an inner journey, with God as the destination, and it can be accomplished. There are other destinations than God, however. The original term for enlightenment, *moksha*, is Sanskrit, translated as "liberation." Liberation from what? From suffering, mortality, pain, the cycle of rebirth, illusion, karma—Eastern spirituality offered many cherished goals as it unfolded over the centuries. Even though *moksha* is considered to be a realistic goal, something that every person should aim for, the frustrating truth is that examples of those who have achieved enlightenment are very scarce. Parallels with the unicorn are uncomfortable.

We'd like to approach the pursuit of enlightenment as the natural path for the brain. For centuries before anyone made the mind-body connection, people didn't know—as we know now—that any experience must involve the brain. You cannot see a toaster or a terrapin without activating the visual cortex. The same holds true if you saw an angel, even in your mind's eye. As far as the neurons in the visual cortex are concerned, an image can be waking or dreaming; it can exist either "in here" or "out there." Nothing visual is possible without stimulating this area of the brain. Nor is it just angels. For God, Satan, the soul, ancestral spirits, or any spiritual experience to exist, your brain must be able to register it, hold on to it, and make sense of it. We aren't speaking of the visual cortex alone. The entire brain is virgin territory for spirituality.

The Brain Wakes Up

One clue that enlightenment is real—and accessible—is staring at us already. We use common phrases all the time that edge right up to it: *Wake up, See the light, Face reality.* These are all pointers to a higher state of awareness. An enlightened person has just gone further. In enlightenment you wake up completely, you see with total clarity, and you face ultimate reality. Your brain therefore is no longer dull and sleepy; instead, it matches your enlightened state, which is alert, vibrant, and creative.

A dramatic shift has occurred, and it's no wonder that in an age of faith, we would put *waking up* in religious terms. In the New Testament, *to see the light* meant "to see the light of God." When Jesus said, "I am the light of the world" (John 8:12), he meant that people would see divinity if they looked at him not as a packet of flesh and blood but as part of God's being. God is the supreme light, and it takes new eyes, the eyes of the soul, to perceive him. Yet any kind of perception, however holy or poetic the terminology, must involve a change in how the brain functions.

When such a shift occurred, you would see everything in a new light, including yourself. Jesus told his disciples not to hide their light under a bushel basket, because they too were part of God. They needed to see themselves with the eyes of the soul and then let the world see how transformed they were. Religions try to patent personal transformation and make it exclusive, but this is a universal process rooted in the mind-body connection. When we say *Face reality*, we mean look at things as they really are, not as illusion. An enlightened person has freed her mind from all illusions and sees reality with perfect clarity. What looks ordinary suddenly becomes divine.

Once the mind wakes up, sees the light, and faces reality, the brain undergoes its own physical changes. Neuroscience cannot fully map these changes, because subjects to test and scan are so few. The whole issue of higher consciousness is being chipped away at, and progress may be very slow. It's nearly impossible to decide if people really see angels, when neuroscientists can't explain how the brain sees anything. As we have pointed out, when you look at the most common object—a table, chair, or book—there is no picture of that object in your brain. So theories of sight, along with the other four senses, remain rudimentary and largely a guessing game.

But the existing evidence for enlightenment, although it has arrived in bits and pieces, is positive. For decades adept Indian yogis have performed remarkable physical feats under the gaze of science. A class of holy men, known as *sadhus*, subject their bodies to extreme conditions as a devotional practice, as well as to gain self-control. Some have been buried underground in a sealed box and survived for days, because they could lower their heart and respiration rates almost to nil. Others survive on meager calories per day or perform exceptional feats of strength. Through specific spiritual practices, yogis and *sadhus* have gained control over their autonomic nervous system; that is, they can consciously alter bodily functions that are generally involuntary.

It would be amazing to witness such extreme control, but it's a limited amazement compared with enlightenment. There the brain adopts a totally new picture of the world, and once the brain changes, the person is filled with wonder and bliss. You undergo a series of *aha!* experiences, and as your brain processes these milestones, you enter a new way of looking at the world. With each *aha!,* an old perception is overturned.

The Aha! *of Enlightenment*

A TRAIN OF INSIGHTS

I am part of everything.
Overturns the belief that you are alone and isolated.

I am cared for.
Overturns the belief that the universe is empty and
   impersonal.

I am fulfilled.
Overturns the belief that life is a struggle.

My life is meaningful to God.
Overturns the belief that God is indifferent (or doesn't
   exist).

I am unbounded, a child of the universe.
Overturns the belief that human beings are an insignificant
   speck in the vastness of creation.

These *aha!s* do not appear all at once. They are part of a process. Because the process is natural and effortless, everyone has moments of awakening. Perception isn't hard to shift. In the movies (and sometimes in real life), a woman exclaims to a man, "Wait a minute. We're not just friends. You're in love with me! How could I have missed it?" This moment of awakening, whether in a movie or in real life, can upend someone's life. But even if it doesn't, the person experiences an internal shift. The mind—accompanied by the brain—stops computing a world based on *we're just friends* in favor of a world in which *you love me* has suddenly entered. Enlightenment follows the same course. Reality A (the secular world) is altered by a moment of insight, shifting your life to different rules, the ones that apply to reality B (in which God is real).

In their yearning for more meaning and fulfillment, people crave reality B. If anyone had a 100 percent guarantee of God's existence, giving up reality A would be a joy and a relief. They would have no more suffering, doubt, or fear of death, no more worry about sin, Hell, and damnation. Religions thrive by feeding our desire to escape the pitfalls of the secular world, however comfortable reality A might be.

The only guarantee that God exists comes from direct experience. You have to feel a divine presence or sense God at work, whatever those phrases mean to you. God plays a relatively small part in the enlightenment process, surprisingly. The greater part is about a shift in perception: waking up, seeing the light, and facing reality. It's a mistake to believe that an enlightened person is a kind of escape artist, a spiritual Houdini who mysteriously frees himself from the illusion of earthly life. The actual purpose of enlightenment is to make the world more real. Unreality comes from thinking that you are isolated and alone. When you see that you are connected to everything in the matrix of life, what could be more real?

There are degrees of enlightenment, and you never know what the next burst of insight will be. There's a potential *aha!* in every situation, if you learn a new way to perceive it. Here's an example from our own lives. At a conference, Deepak met a noted neuroscientist who mentioned that she was more comfortable in the world of birds than in the world of people. What could such a statement mean? It didn't appear to be delusional. This woman knew her neuroscience very well; she was intelligent and articulate.

What her experience came down to was something like that of the horse whisperer: tuning in to the nervous systems of other creatures. A decade ago the claim would have seemed flaky. How can someone think like a dog, the way Cesar Millan does, or like a horse, the way Monty Roberts, the original horse whisperer, does? The answer is sensitivity and empathy. Being self-aware, we can already extend our awareness to how other people feel. There's no mystery to feeling someone else's joy or pain.

It seems that we can do the same with animals, and the proof is that you can train horses and dogs almost effortlessly if you whisper in their language, without using whips, muzzles, or mistreatment. When you know how an animal's nervous system sees the world, you don't have to "break" the animal. You can change animal behavior most easily by following the natural course of an animal's brain.

In the case of the bird lady, the proof of her attunement is that several types of wild birds are comfortable sitting on her shoulder and eating out of her hand. Does that make her the heir to Saint Francis of Assisi, who is depicted in just this way? Yes, in a sense. A saint's ability to see the whole of creation as part of God brings a sense of empathy with all living things. A shift occurs in the saint's nervous system that expresses what the mind now accepts: *I am at peace with the world and all living things. I am not here to harm them.*

Is it so amazing that other creatures know when we come in peace? Our pets sense who to growl at and who to approach for a pat on the head. The human nervous system has commonality with that of other creatures. It sounds very dry to put this in such an analytical way, but the truth is quite beautiful when you see a bird alight in the palm of your hand.

Deepak recounted his meeting with the bird lady, but it wasn't yet an *aha!* experience. Rudy triggered the *aha!* when Deepak asked a far-out question: Since human DNA is 65 percent the same as a banana's, can we empathize with bananas, or communicate with them? (He had in mind some famous experiments by Cleve Backster, who connected houseplants to sensitive electrical sensors and found that plants displayed changes in their electrical field, as measured with a kind of polygraph or lie detector, when their owners argued or exhibited high stress. In the most startling finding, plants displayed the greatest electrical excitation when their owners thought about chopping them down.)

Rudy replied that when we taste the sweetness of a banana, the receptors on our tongue are connected to the banana's sugar, so in a sense we are participating in its reality at the chemical level. The banana also provides us with proteins that bind receptors that are similar to ours. So we experience a "molecular" form of communication. In the same way, when you digest a banana, its energy gets transformed into your energy, which is a link even more intimate than communication. When you analyze the total DNA in a human, it turns out that more than 90 percent comes from the bacteria that inhabit our bodies in a mutually dependent (symbiotic) manner. Much of our own human DNA is similar to bacterial DNA. And the major organelles that provide us with energy, called mitochondria, are actually bacterial cells that were integrated into our cells for this purpose. Therefore we are genetically woven into the web of life. It forms one matrix of energy, genes, and coded chemical information. No part is separate or isolated. That's the *aha!*. More and more people are getting the same *aha!*, as evidenced by the rise of the modern ecology movement. Humans are beginning to abandon the illusion that the Earth is ours to manipulate and damage at will, with no dire consequences for us. But even without data about vanishing ozone and rising ocean temperatures, the ancient sages and seers of India, as part of their journey to enlightenment, arrived at the same insight when they declared "the world is in you." Ecology weaves together every activity that supports life, whether it takes place in our cells or in a banana's.

Where's the Proof?

The skeptical viewpoint holds that when a person believes in God, the brain, being capable of creating illusions, fools itself into believing and adopts all the trappings of spirituality. To a skeptic, simple material reality (*This rock is hard. That's what makes it real.*) is the only kind. All spiritual experience therefore has to be unreal, never mind that blanket doubt applies to Jesus, Buddha, Lao-Tzu, and countless saints and sages revered for thousands of years. It's all rubbish to the confirmed skeptic. Richard Dawkins, the British ethnologist and science writer who presents himself as a professional atheist, wrote a book for young people, *The Magic of Reality*, which addresses the whole issue of what is real. He informs the reader that if we want to know what's real, we use our five senses, and when things are too big and far away (e.g., distant galaxies) or too small (e.g., brain cells and bacteria), we augment our senses with devices like telescopes and microscopes. One anticipates that Dawkins will add a caveat that the five senses aren't always reliable, as when our eyes tell us that the sun rises in the sky in the morning and sets at twilight, but he offers no such caveat.

In Dawkins's scheme, nothing that we know emotionally or intuitively is valid, and the most fraudulent belief of all is "the God delusion." (By no means does he speak for all scientists. According to some polls, scientists believe in God and even attend religious services more frequently than the general population.)

The rift between materialism and spirituality—facts versus faith—is centuries old by now, and yet the brain may heal it. Solid research about meditation has confirmed that the brain can adapt to spiritual experiences. In Tibetan Buddhist monks who devote their lives to spiritual practice, the prefrontal cortex displays heightened activity; the gamma wave activity in their brains has twice the frequency of normal people's. Remarkable things are happening in the monks' neocortex that brain researchers have never seen before.

So shrugging off spirituality as mere self-delusion or superstition is contradicted by science itself.

Skepticism is not the problem; in fact, it's a red herring. The real problem is the mismatch between modern life and the spiritual journey. Countless people yearn to experience God. Spending a lifetime on the inward path may be deeply rewarding, but few people are lifelong seekers in the traditional sense. Spiritual needs have changed since the age of faith. God has been put on the shelf. As for enlightenment, it's too difficult, too far away and improbable. The brain can help us here, too. Let's redefine the state of enlightenment in updated terms. Let's call it the highest state of fulfillment. What would such a state look like?

> Life would be less of a struggle.
> Desires would be achieved more easily.
> There would be less pain and suffering.
> Insight and intuition would become more powerful.
> The numinous world of God and the soul would be a real
>   experience.
> Your existence would feel deeply meaningful.

These goals give us a realistic process that can move ahead by degrees. Enlightenment is about total transformation, but not instant transformation. The brain undergoes its physical shifts as you, its user and leader, reach new stages of personal change. Here is what to look for. Far from being exotic, these are aspects of your own awareness at this moment; all you need to do is expand them.

SEVEN DEGREES OF ENLIGHTENMENT

Inner calm and **detachment** increase—you can be centered in the midst of outer activity.

Feeling connected grows—you feel less alone, more bonded with others.

Empathy deepens—you can sense what others are feeling, and you care about them.

Clarity dawns—you are less confused and conflicted.

Awareness becomes more **acute**—you get better at knowing what's real and who is genuine.

Truth reveals itself—you no longer buy into conventional beliefs and prejudices. You are less swayed by outside opinions.

Bliss grows in your life—you love more deeply.

You don't try to achieve these separate aspects of expanded awareness with a head-on attack. Rather, each one appears in its own time following its own rhythm. Nothing needs forcing. One person will notice an increase in bliss earlier—and more easily—than they feel more clarity, while for a second person the reverse is true. As enlightenment unfolds, it follows your nature, and we're all made differently.

The key is to *want* enlightenment in the first place, which is tied to the whole issue of transformation.

If you want to transform yourself, which is what enlightenment is all about, what does your brain need to do? If it can shift as easily as it is shifting right this minute, there is no serious obstacle. The millions of people yearning for personal transformation already have it in hand, actually. Their brain is transforming itself constantly. Just as you cannot step into a river in the same place twice, you can't step into the brain in the same place twice. Both are flowing. The brain is a process, not a thing; a verb, not a noun.

Our biggest mistake is to believe that transformation is uncommonly hard. Imagine an experience in your past that flooded over you and left you feeling changed. The experience could be positive, like falling in love or getting a big promotion; or it could be negative, like losing your job or getting a divorce. In either case, the effect on your brain is both short-term and long-term. This applies to memory, since you have specific brain regions for short-term and long-term memory, but the effect goes much further. Overwhelming experiences change your sense of self, your expectations, your fears and wishes for the future, your metabolism, your blood pressure, your sensitivity to stress, and anything else monitored by the central nervous system. Overwhelming experiences transform you.

A good movie is enough to cause powerful shifts in the nervous system. Hollywood blockbusters compete to explode the audience's sense of reality and offer new vicarious thrills. Spider-Man swings through the man-made canyons of New York City on sticky ropes, while Luke Skywalker jockeys his lightship to enter the Death Star, and dozens of other dazzling effects exist to transform the brain.

When you walk out of a movie, its effect stays with you; the glow is more than just temporary. Kissing the girl in your own mind, defeating the villain, marching with the conquering heroes—viewed from the level of the neuron, none of these experiences are unreal. They are real because your brain has been altered. A movie is a transformation machine, and so is life itself. Once you accept that transformation is a natural process—one that every cell participates in—enlightenment is no longer out of reach.

Of course, winning the girl in a movie didn't happen in real life. Your brain is fooled for a while, but you aren't. You bring yourself back to reality (where love and romance lead to the knotty problems of relationships). This is the key. Bringing your attention back to what is real can become a spiritual practice, known as *mindfulness*. Mindfulness can be made into a way of life, and when it is, transformation becomes a way of life too, as natural and unforced as anyone could wish.

The Mindful Way

What are you aware of right this minute? Perhaps you don't have your attention on anything beyond the words on this page. But as soon as the question is asked, "What are you aware of?" your perception wakes up. You notice all kinds of things: your mood, the comfort or discomfort of your body, the temperature in the room, and the light radiating inside it. This shift, which calls attention to reality, is mindfulness.

You can bring your awareness back to reality anytime you want to. You don't have to force anything; you don't need superhuman willpower. But mindfulness does feel different from ordinary awareness. Our awareness is normally focused on a specific object or task. That's how we train our brains—to see the thing before our eyes but not the background, which is awareness itself. The background we take for granted—until we are shocked back into awareness of it. Imagine that you are on a date with someone who seems very attentive. He (or she) can't take his eyes off you. He's hanging on to every word. Naturally, you get lost in the pleasure of this feeling. But then he says, "I'm sorry, but do you know you have some spinach between your teeth?"

At that moment your awareness shifts. You have been shocked out of your pleasant illusion. Being brought back to reality doesn't have to be unpleasant. Imagine that you are about to meet a VIP, and you're feeling nervous and anxious. The moment before you shake hands, however, somebody leans over and whispers in your ear, "Mister Big has heard great things about you. He already wants to give you the contract." Another kind of shift occurs: you switch from an anxious state into a more confident one. Mindfulness is about the ability to do so.

The ability comes naturally. A few words whispered in your ear can trigger dramatic, instantaneous change. At the level of hormones, we know part of the answer, but we are far from knowing how the brain shifts its reality at the flick of a switch. But there is clearly a difference between owning this ability and letting our brain own it. Mindfulness makes the difference. Instead of having other people shock you back to reality—whether the shock is pleasant or unpleasant—you bring yourself back. To define *mindfulness* as "awareness of awareness" fits, but it sounds, to us, arcane; the simpler explanation is that you can return to reality anytime you want.

Unfortunately, we have all surrendered some of this ability. Certain areas of our life are safe to pay attention to, while others are off limits. Women typically like to discuss their feelings, for example, and complain that men don't or won't or can't. Men typically are more comfortable focusing on work, sports, and various projects—almost anything that doesn't touch an emotional sore spot. But in Eastern spiritual traditions, there's a vast field that most Westerners barely think about: awareness of awareness. The term for this in Buddhism is mindfulness.

Whenever you check in on yourself, you are being mindful. Before a date or job interview, you might check to see how nervous you feel. During childbirth, as the doctor asks, "How are you doing?" a woman monitors if her pain is getting too great. In this very basic kind of mindfulness, you are looking at moods, emotions, physical sensations—all the things that fill the mind. What if you took away the contents of your mind? Would you face a frightening cold emptiness? No. A great painter might wake up one day to discover that all his paintings had been stolen, but he would still have something invisible and far more precious than any masterpiece: the ability to create new ones.

Mindfulness is like that, a state of creative potential. Once you take away the contents of the mind, you have the most potential, because you are in a state of complete self-awareness. (Once a music lover came to the renowned spiritual teacher J. Krishnamurti and exclaimed rapturously about how beautiful a concert had been. Krishnamurti replied astutely, "Beautiful, yes. But are you using music to distract you from yourself?") True mindfulness is a way of checking in on how self-aware you are. As you know by now, super brain depends upon a growing self-awareness, so being mindful is crucial. It's a way of life.

People who aren't mindful can seem at once oblivious and self-involved. They are too self-centered to connect with other people; they lack sensitivity in many kinds of social situations. The contrast between being self-centered and being mindful is quite striking, so let's look at the difference. Both states are produced in the neocortex, yet they don't feel the same. Being self-centered almost demands that you indulge in illusions, since everything revolves around your image. We aren't judging against being self-centered; it's the perspective that consumer society trains us all to have—it goads us to buy things that will make us better looking, younger, hipper, more entertained, and momentarily distracted.

Self-centered: Your thoughts and actions are dominated by I, me, mine. You focus on specific things that you can achieve or possess—you set goals and meet them. The ego feels in control. Your choices lead to predictable results. The world "out there" is organized through rules and laws. External forces are powerful but can be reined in and managed.

TYPICAL THOUGHTS

I know what I'm doing.
I make my own decisions.
The situation is under control.
I trust myself.
If I need help, I know where to turn.
I'm good at what I do.
I like a challenge.
People can depend on me.
I'm building a good life.

Mindful: Your mind is reflective. It turns inward to monitor your sense of well-being. Self-knowledge is the most important goal. You don't identify with the things you can own. You value and often rely upon insight and intuition more than logic and reason. Empathy comes naturally. Wisdom dawns.

TYPICAL THOUGHTS

This choice feels right, that other one doesn't.
I'm tuning in to the situation.
I know just how others feel.
I see both sides of the issue.
Answers just come to me.
Sometimes I feel inspired, and those are the best times.
I feel like a part of humankind. No one is alien to me.
I feel liberated.

The mindful state is just as natural as any other. When we overlook it, we create unnecessary problems.

Some years ago, for example, Rudy was in a rush to complete some experiments, before making a seven o'clock flight out of Boston—he was scheduled to give the opening lecture at a major international conference. Caught in the city's notorious rush-hour traffic, however, he ran out of luck and missed his flight. Standby was uncertain, but if he didn't catch the last flight out, he'd have to suffer the embarrassment of being a no-show. Rudy became anxious and angry. Screaming at the counter agent would do no good, but he was tempted. Completely unaware of it, he was identifying with the intensely negative feelings that his brain was producing.

Of course, most people would consider these feelings completely natural in the situation. But the healthier alternative would have been for Rudy to experience his frustration for a limited amount of time and then become mindful. Standing back, he could have observed how missing the flight triggered his instinctive/emotional brain, producing a full-fledged stress reaction in his body. Without mindfulness, the stress would run its course over a longer period of time, and unfortunately, as the years pass, our bodies become more easily stressed and recover more slowly from small incidents—letting the stress reaction run its course is not healthy. In the end, stress breeds stress.

By becoming the active observer of the negative feelings evoked in his brain, Rudy could have more proactively dealt with the situation and learned from it. Most important of all, he would not have been the victim of the reactive mind. This isolated incident summarizes the advantages of mindfulness:

> You can handle stress better.
> You free yourself from negative reactions.
> Impulse control becomes easier.
> You open a space for making better choices.
> You can take responsibility for your emotions instead of
> blaming others.
> You can live from a place that is more centered and calm.

How can you cultivate mindfulness? The short answer is meditation. When you close your eyes and go inward, even for a few minutes, your brain gets a chance to reset itself. You have no need to try to become centered. The brain is designed to return to a balanced, unexcited state when given the chance. At the same time, when you meditate, a change occurs in your sense of self. Instead of identifying with moods, feelings, and sensations, you put your attention on quietness, and as soon as this happens, the stress that was agitating you is no longer as sticky. When you stop identifying with it, stress has a much harder time taking hold.

The practice of meditation is not as alien to most people as it was three or four decades ago, and there are many advanced kinds. But starting with the most basic techniques often produces a startling contrast. Sit in a quiet place and close your eyes. Make sure you have no distractions; turn the lights to dim.

As you sit, take a few deep breaths, letting your body relax as much as it wants to. Now quietly notice your breath going in and out. Easily let your attention follow your breathing, as you would if you were sitting in an easy chair listening to a gentle summer breeze. Don't force yourself to pay attention. If your thoughts wander—which always happens—gently bring your awareness back to your breath. If you wish, after five minutes, put your attention on your heart, and let it rest there for another five minutes. Either way, you are learning something new: how it feels to be in a mindful state.

To go even deeper, you might use a simple mantra. Mantras have the benefit of taking the mind to a subtler level. Sit quietly and take a few deep sighs, and when you feel settled, think the mantra *Om shanti*. Repeat it as you feel like it, but don't force a rhythm; this isn't mental chanting. Don't follow your breath. Just repeat the mantra whenever you notice that your attention has wandered away from it. There's no need to think it quietly—it will grow quiet on its own—but certainly don't think it loudly. Do this for ten to twenty minutes.

Newcomers will naturally ask how they are to know if meditation is working. If you lead an active life and expend too much energy, your body will so desperately need rest that you will spend many meditations falling asleep. This isn't a failure; your brain is taking what it needs the most. But especially if you meditate in the morning, before you start your day, you will experience the quietness of awareness looking at itself. After ten to twenty minutes, you will notice how easy, relaxed, and comfortable it feels to be centered.

We said that meditation was the short answer, because there's the entire rest of the day to consider. How can you be mindful outside meditation? The principle here will be familiar: change without force. Staying centered and mindful all day isn't something you can force. But you can gently favor the behavior of a mindful person:

> Don't project your feelings onto others.
> Don't participate in negativity.
> When you feel stress in the air, walk away.
> Don't put your attention on anger and fear.
> If you have a negative reaction, let it run for a little while; then as soon as you can, step back, take a few deep breaths, and observe your reaction without indulging it.
> When you are having a reaction, don't make any decisions until later, when you are once again centered.
> In your relationships, don't use arguments to vent your resentments. Discuss your issues when you both feel calm and reasonable. This is an easy way to avoid delivering unnecessary wounds in the heat of the moment.

In practical terms, being mindful is self-monitoring without casting blame or judgment. When you don't monitor yourself, you can fall prey to a wide range of difficulties. "I don't know why I did that" is the most common complaint when people aren't mindful, along with "I was out of control." In the aftermath of impulsive reactions, they feel remorse and regret.

From the brain's perspective, when you self-monitor, you are introducing a higher state of balance. The primitive reactions of the brain are rarely appropriate in modern life. They persist as if humans still needed to fight predators, fend off raiding tribes, and run away from threats. In the course of evolution, the higher brain has evolved to introduce a second response, which is more suited to the situation's actual level of threat. But for most people most of the time, there is no threat at all. You don't need the lower brain's primal reactions, even though they will keep springing up—they are biologically wired in.

When the lower brain acts inappropriately, you can defuse it by reminding yourself of reality: you are not being threatened. That awareness alone is enough to reduce many kinds of stress reactions. Mindfulness goes further, however. After spending some time meditating, you will find a higher balance—you will start to identify with a peaceful state of restful alertness. That will open the doorway to the sort of spiritual experience that would be out of reach otherwise. A lovely passage from the *Mandukya Upanishad* of ancient India describes how necessary the mindful state is:

Like two birds perched in the same tree, who are intimate friends, the ego and the self dwell in the same body. The first bird eats the sweet and sour fruits of life, while the other bird silently looks on.

As you become more mindful, both sides of your consciousness will be recognized, and then they can become the intimate friends described in this passage. Ego, the restless, active *I*, no longer must act on its drives and desires. You learn that the self, the other half of your nature, is content simply to be. There is immense fulfillment in finding that you are enough within yourself, needing no outside stimulation to make you happy. We call this fusion the true self.

SUPER BRAIN SOLUTIONS

MAKING GOD REAL

We want to shed light on the age-old dilemma of whether God exists. Mindfulness can help here, because when it comes to matters of faith and hope, awareness is crucial. There's a huge gap between *I hope, I believe,* and *I know.* This is true of everything that happens in your awareness, not just with God. Is your spouse cheating? Can you handle being made a supervisor at work? Are your kids going to take drugs? In one way or another, the answers lie in the vicinity of three choices: you hope, you believe, or you know that you have the right answer. But since God is the toughest of these choices, we'll focus on him (or her).

In spiritual matters, faith is supposed to be the answer, but its power seems limited. Almost everyone has made a personal decision about God. We say God doesn't exist or he does. But our decision is usually shaky and always personal. "God doesn't exist for me, at least I think he doesn't" would be more accurate. How can you tell if deep spiritual questions have an answer you can trust? Does the same God apply to everyone?

As children, we all asked the most basic spiritual questions. They came naturally: *Does God look after us? Where did Grandma go after she died?* Children are too young to understand that their parents are just as confused in these matters as they are. Children get reassuring answers, and for a time they suffice. If told that Grandma went to Heaven to be with Grandpa, a child will sleep better and feel less sad. When you grow up, however, the questions return. And thus you discover that your parents, however well intentioned, never showed you the way to find answers, not just about God but about love, trust, your life's purpose, and the deeper meaning to existence.

In all these cases you either hope, believe, or know what the answer is: "I hope he loves me." "I believe my spouse is faithful." "I know this marriage is solid." These statements are very different, and we find ourselves awash in confusion because we don't differentiate between *I hope, I believe,* and *I know,* as if they were the same. We just wish they were. We shy away from seeing where things really stand.

Reality is a spiritual goal as much as a psychological one. The spiritual path takes you from a state of uncertainty (*I hope*), to a somewhat firmer state of security (*I believe*), and eventually to true understanding (*I know*). It doesn't matter whether the specific issue is about relationships, God, or the soul, about the higher self, Heaven, or the domain of departed spirits. The path begins with hope, grows stronger with faith, and becomes solid with knowing.

In these skeptical times, many critics try to undermine this progression. They claim that you cannot know God, the soul, unconditional love, the afterlife, and a whole host of other profound things. But the skeptic scorns the path without having set foot on it. If you look back at your past, you'll see that you have already made this journey, many times in fact. As a child you hoped you would be a grown-up. In your twenties you believed that it was possible. Now you know you are an adult. You hoped someone would love you; you believed in time that somebody did; and now you know that they do.

If this natural progression hasn't happened, something has gone wrong, because the unfoldment of life is designed to lead from desire to fulfillment. Of course, we all know the pitfalls. You can say to yourself "I know I'll make it big," when in fact you are only hoping. Getting a divorce may mean that you didn't know if someone truly loved you. Children who grow up resenting their parents usually don't know who to trust. A hundred other examples of broken dreams and lost promises could be offered. But far more often the

progression works. Desires are the things that drive life toward fulfillment. What you hope for, one day you will know.

Certain aspects of mindfulness come into play here, and they
seem to be universal. They are important for anyone who doesn't
want to be trapped in futile wish fulfillment and faith that isn't
based in reality. You can only trust what you truly know.

How Do You Know?

When you truly know something, the following things apply:

You didn't accept other people's opinions. You found out on
your own.

You didn't give up too soon. You kept exploring despite blind
alleys and false starts.

You trusted that you had the determination and curiosity to
find out the truth. Half-truths left you dissatisfied.

What you truly know grew from the inside. It made you
a different person, as different as two people when one of
them has fallen deeply in love and the other hasn't.

You trusted the process and didn't let fear or discouragement
impede it.

You paid attention to your emotions. The right path feels a
certain way, satisfying and clear; uncertainty is queasy and
gives off a bad smell.

You went beyond logic into those areas where intuition, insight, and wisdom actually count. They became real for you.

What makes this scenario universal is that the same process applies to the Buddha seeking enlightenment or to any young person
learning how to be in a relationship or finding her purpose in life. By

dividing the process into its components, the huge questions about life, love, God, and the soul become manageable.

You can work on one ingredient at a time. Are you prone to accepting secondhand opinions? Do you distrust your own decisions? Is love too painful and confusing to explore deeply? These aren't impossible obstacles. They are part of you, and therefore nothing can be nearer or more intimate. But let's be even more specific. Think of a problem that you want to solve, something deeply meaningful to you. It can be as philosophical as "What is my purpose in life?" or as spiritual as "Does God love me?" It can be about a relationship or even a problem at work. Pick something hard to solve, where you feel doubt, resistance, and stuckness. You keep hoping to find an answer, but so far you haven't been able to.

Whatever you choose, finding an answer that you can trust involves taking certain steps.

MOVING FROM HOPE TO FAITH TO KNOWLEDGE

Step 1: Realize that your life is meant to progress.

Step 2: Reflect on how good it is to truly know something rather than just hoping and believing. Don't settle for less.

Step 3: Write down your dilemma. Make three separate lists, for the things you hope are true, the things you believe are true, and the things you know are true.

Step 4: Ask yourself why you know the things you know.

Step 5: Apply what you know to those areas where you have doubts, where only hope and belief exist today.

As applied to God or the soul, we've taken an issue that most people consider mystical, requiring a leap of faith, and broken it down. The brain likes to work coherently and methodically, even when it comes to spirituality. The first two steps are psychological preparation; the last three ask you to clear your mind and open the way for knowledge to enter. Let's apply the steps to God now.

Step 1: Realize that your life is meant to progress.

In spiritual terms, progressing means wanting to come to terms with God; you feel deserving and know that the benefits of a loving deity would be good in your life. This is the opposite of Pascal's famous wager, which says that you might as well bet that God exists, because if you disbelieve and God turns out to be real, you might wind up in Hell. The problem is that Pascal's wager is based on fear and doubt. Neither is a good motivator for spiritual growth. Instead, think of how fulfilling it will be to know whether God is real, not how bad it will be if you wind up on the wrong side of a bet.

Step 2: Reflect on how good it is to truly know something.

Here you put your mind on finding God as a valid experience, not a trial of faith. When you sense doubts and misgivings—which we all have around God—don't shrug them off. Open a space for the possibility that all the charges leveled against God might not be the whole story. Despite every woe that human life is heir to, including the very worst that are leveled against a loving God (genocides, wars, atomic weapons, despots, crime, disease, and death), the issue isn't settled by any means. A loving God might still exist who allows humans to commit mistakes and learn at their own pace. Don't jump to any conclusions, however. Adopt the attitude that you can solve the problems of violence, guilt, shame, anxiety, and prejudice—the roots of global problems—in your own life. Undertaking personal growth is far better than bemoaning the perennial state of human suffering.

Step 3: Write down your dilemma, making three separate lists, for the things you hope are true, the things you believe are true, and the things you know are true.

The point here is to avoid generalizations and received opinion. Most people make a blanket judgment for or against God, then hedge their bets according to the situation at hand. (As the saying goes, there are no atheists in the foxholes. There are probably also few devotees praying in a singles bar after midnight.) By listing your hopes, beliefs, and real knowledge, you will surprise yourself. Spiritual issues are fascinating once you decide to pay attention to them. As a secondary benefit, you will sharpen and clarify your thinking, which aids your higher brain. Thinking is a skill organized in the neocortex, and this includes thinking about God.

So be frank. Do you secretly believe that God punishes sinners, or do you hope that he doesn't? If both are true, note this in two lists, the one for hope and the one for beliefs. Do you think you've witnessed an act of grace or forgiveness? If so, put this down as something you know. As a beginning to spiritual exploration, this exercise is very revealing. Take time over your lists, and when you finish them, put them away where you can go back and consult them in the future—that's a good way to see how well, and realistically, you are progressing.

Step 4: Ask yourself why you know the things you know.

The blunt phrase "I know what I know" papers over a lot of complexity. Most people let their beliefs settle in place without considering where they came from. Do you believe in God (if you do) because your parents told you to, or you accepted the lessons in Sunday school? Perhaps your belief rests upon a desperate hope that the man upstairs is watching you; but to be realistic, you don't actually know if God is a man, and upstairs could be anywhere, nowhere, or everywhere in creation.

To actually know about God, it's certainly best to have personal experiences, but these cover a wider range than you might suppose.

Have you felt a divine or luminous presence?

Have you felt loved in an all-embracing way?

Have you felt a sudden surge of bliss or joy that you couldn't attach a cause to?

Have you felt safe and cared for, as if your existence is accepted by the universe?

Do you experience times of deep inner calm, strength, or knowingness?

As you can see, the word *God* doesn't have to be connected to experiences of expanded awareness, which is what you want your brain to register and remember. In polls, almost a majority of people say that they have seen a light around someone else, and many people have experienced healing or the power of positive thinking. The issue isn't whether you have met God; the issue is your actual experience that might direct your mind to a world that reaches beyond the material.

As you consider the kind of experience that you know to be real in your own life, you can also think about scriptures and the people who wrote them. If you know that you enjoy reading the Bible or the poems of Rumi, if you have felt peace around a spiritual person or in a holy place, then you know something to be true. By paying attention and making such experiences significant, you go a long way toward finding your place in the spiritual matrix, just as you have a place in the matrix of life.

Step 5: Apply what you know to those areas where you have doubts.

If you've taken the first four steps, you should have a good mind map of your present state of hope, belief, and knowledge. This alone is very helpful, since it gives you a basis for any signs of change. Change requires intention, and if you tell your brain that you intend to look for God, your powers of perception start to increase. (Doesn't this happen when you decide that you want to look for a love interest? Suddenly you see the people around in a different, sharper light—strangers turn into prospects for romance, or not.)

God likes to be engaged, which is to say, taking an interest in spiritual growth isn't passive. You must open yourself up to walking the walk, spiritually speaking. Contrary to common belief, this doesn't mean making a New Year's resolution to go back to church (not that we advise against it by any means) or deciding overnight to be saintly and devout. Those are points of arrival more than points of departure. The core issue is how to act in such a way that the possibility of God becomes real.

We call such activity "subtle actions," because they take place inside. Consider the following subtle actions and how you could adapt to them.

ACTING AS IF GOD MIGHT BE REAL

Meditate.

Have an open mind about spirituality. Examine any tendency to be reflexively skeptical and closed off.

See the good in people. Stop gossiping, blaming, and taking silent petty pleasure when bad things happen to people you don't like.

Read uplifting poetry and scriptures from many sources.

Look into the lives of saints, sages, and seers from spiritual traditions East and West.

When in distress, ask to have your anxiety removed and your burden lessened.

Leave room for unexpected solutions. Don't force the issue or fall back on the need to control.

Fully experience joy each day. Do this even if it's only gazing at a blue sky or smelling a rose.

Spend time around children and absorb their spontaneous exuberance for life.

Be of service to someone in need.

Consider the possibility of forgiveness somewhere in your life where it will really make a difference.

Reflect on gratitude and the things you are grateful for.

When you feel anger, envy, or resentment in a situation, stand back, take a deep breath, and see if you can let it go. If not, at least postpone your negative reaction until a later time.

Be generous of spirit.

Expect the best unless you have evidence that something needs help, improvement, or criticism.

Find a way to enjoy your existence. Address the serious obstacles that prevent your enjoyment.

Do what you know to be good. Avoid what you know to be bad.

Find a path to personal fulfillment, however you define that word.

This list gives you some specificity so that God doesn't become a vague wash of emotion or a topic to postpone until a crisis looms. We have avoided religiosity, not because we are arguing against any faith but because the goal is different here. You want to gently train your brain to notice and value a new reality. Whether to participate in such a reality is your choice. Just be aware that if you want to attune yourself to the vast matrix of spiritual experience, your brain is ready to adapt.

In a sense, the simplest advice we've heard about God is also the most profound. At least once a day, let go and let one situation be handled by God or your soul or whatever agent of higher wisdom you choose. See if your life can take care of itself. For in the end, it isn't the man upstairs—or an entire pantheon of gods—who direct the course of life. Life evolves within itself, and God is only a label we apply to unseen powers that exist, waiting to emerge, from within ourselves. When you read the following couplet of the great Bengali poet Rabindranath Tagore, be mindful of how you feel:

Listen, my heart, to the whispering of the world.
That is how it makes love to you.

Or this one:

How the desert yearns for the love of just one blade of grass!
The grass shakes her head, laughs, and flies away.

If you feel the tenderness of the first couplet and the mystery of the second, a place inside you has been touched as surely as if God touched you. There is no difference, except that the experiences grow until the divine is real for you. This is your privilege. There is no need for it to be real for anyone else.

THE
REALITY ILLUSION

We cannot fully explore the brain without addressing its deepest mystery. You are immersed in it every second of your life. Imagine that you are on vacation gazing at the Grand Canyon. Photons of sunlight glancing off the cliffs make contact with your retina and stream into your brain. There the visual cortex is activated through chemical and electrical activity, which comes down to electrons bumping into other electrons. But you aren't aware of this stormy, minuscule process. Instead, you see vibrant color and form; the awe-inspiring chasm appears before you, and you hear the whistling wind rush out of the canyon and feel the hot desert sun on your skin.

Something almost indescribable is happening here, because not a single quality of this experience is present in your brain. The Grand Canyon glows a brilliant red, but no matter how hard you search, you won't find a spot of red in your neurons. The same holds true for the other four senses. Feeling the wind in your face, you won't find a breeze in your brain, and its temperature of 98.6 degrees Fahrenheit won't change, whether you are in the Sahara or in the Arctic. Electrons bump into electrons, that's all. Since electrons don't see, touch, hear, taste, and smell, your brain doesn't either.

As mysteries go, this one is a stumper. Your consciousness of the world around you can't be explained if you insist upon a materialistic model. Yet the model based on electrical and chemical reactions, which are materialistic, is exactly what the field of neuroscience keeps pursuing. A flood of new data piles up about the brain's physical operation, creating tremendous excitement. It would help if we knew, with complete certainty, how the mind-brain connection produces the world we see, hear, and touch.

Once when Deepak was giving a talk on the subject of higher consciousness, a skeptical questioner stood up in the audience. "I'm a scientist," he introduced himself, "and this is all smoke and mirrors. Where is God? You can't produce any evidence that he exists. Enlightenment is probably just self-delusion. You have no proof that supernatural things are real." Without pausing to consider, Deepak replied, "You have no proof that *natural* things are real." Which is true. Mountains, trees, and clouds look real enough, but without having the slightest idea how the five senses arose from electrons bumping into electrons, there is no proof that the physical world matches our mental representation of it.

Is a tree hard? Not to the termites boring into it. Is the sky blue? Not to the multitudes of creatures that are color blind. Research has discovered a peculiar trait in crows; they recognize individual human faces and will react when that same face reappears a few days or even weeks later. But a trait that seems so human must have a very different use in the bird world, one we can only imagine, since our nervous systems are tuned only to our reality, not a bird's.

Every one of the five senses can be twisted to deliver a completely different picture of the world. If by *picture* we mean the sight, sound, smell, taste, and texture of things, a troubling conclusion looms. Apart from the very unreliable picture running inside the brain, we have no proof that reality is anything like what we see.

Einstein put it another way when he said that the most incredible thing isn't the existence of the universe but our awareness of its existence. Here is an everyday miracle, and the more you delve into it, the more wondrous it becomes. Consciousness deserves to be called *the* hard problem, a phrase popularized by David Chalmers, a specialist in the philosophy of mind.

We feel that the hard problem becomes much easier when we give consciousness a primary role instead of making it secondary to the brain. We've already shown that you—meaning your mind—are the user of the brain. If you are telling your brain what to do, it isn't a huge leap to say that the mind comes first and the brain second. We've also called you a reality maker. It would close the circle if you are not just reshaping your brain at every moment, not just causing chemicals to fire in the brain, but are actively creating everything in the brain. This is a more radical role for the mind, but far-seeing cognitive scientists and philosophers have taken such a position—it turns out to have many surprising advantages.

The hard problem is abstract, but none of us can afford to leave it to professional thinkers. The best and the worst of what will happen to you today—and everything in between—is the fruit of your awareness. You spend every day adding to the same project, one that lasts a lifetime. Let's call this project "building a self." Everyone has the right to feel unique, but the input for building your self consists of the positive and negative messages that register in your awareness, beginning with things that are painful and things that are pleasurable. The building blocks of the self are made of "mind stuff," so it's not true to say that you *have* consciousness, the way you have a kidney or an epidermis; you *are* consciousness. A fully formed human adult is like a walking universe of thoughts, desires, drives, fears, and preferences accumulated over the years.

The good news is that your brain, which registers and stores all your experiences, gives clear signals of what needs to be changed whenever there is imbalance, dis-ease, and a breakdown in the smooth partnership of mind and body. We can divide the most telltale signals into positive and negative categories.

Building a Self

How many of the following apply to you today?

POSITIVE SIGNALS

Inner calm and contentment
Curiosity
Sense of openness
Feeling of safety
Purposefulness, dedication
Feeling of being accepted and loved
Freshness, physical and mental
Self-confidence
Sense of worth
Alert self-awareness
Absence of stress
Engagement, commitment

NEGATIVE SIGNALS

Inner conflict
Boredom
Fatigue, physical or mental
Depression or anxiety
Anger, hostility, critical attitude toward self and others
Confusion about your purpose
Feeling of unsafeness, insecurity
Hypervigilance, alertness to constant threats
Stress
Sense of low self-worth
Confusion, doubt
Apathy

No matter what stage of life you find yourself in, all the way back to very early childhood, your brain is sending these signals, playing them off against each other without stopping and so contributing to your development of self.

Society guides the building of a self, but each person creates a distinct *I* within the framework. How this is done is complex and little understood. We are expected to create ourselves instinctively. We feel our way through thousands of situations, and the net result is a jury-rigged construct. We took two or three decades to build it and yet none of us really knows how we arrived at the self we inhabit. The whole process needs to be improved. Since everything that creates a self happens in consciousness, you now have a personal reason to solve the hard problem. Some thorny arguments lie ahead, but the end result will be a leap in your well-being.

Ghosts Inside the Atom

From the time of Sir Isaac Newton, physics has been based on the commonsense belief that the physical world is solid and stable. Therefore, reality starts "out there." It's a given. Einstein called this belief his religion. Once when he was walking at twilight with another great quantum physicist, Niels Bohr, the two were talking about the problem of reality. It hadn't been a problem for science until the quantum era, at which point the tiny solid objects known as atoms and molecules began to vanish. They turned into whirling clouds of energy, and even those clouds were elusive. Particles like photons and electrons didn't have a fixed place in space, for example, but instead obeyed laws of probability.

Quantum mechanics holds that nothing is fixed or certain. There is an infinitesimal chance, for example, that gravity won't cause an apple to fall from a tree but will make it move sideways or upward instead, although such anomalies apply not to apples—the chance of an apple not falling is almost infinitely remote—but to subatomic particles. Their behavior is so strange that it gave rise to an aphorism from Werner Heisenberg, the creator of the Uncertainty Principle: "Not only is the Universe stranger than we think, it is stranger than we can think."

To the end of his life, Einstein was uneasy with such strangeness. One particular disagreement had to do with the observer. Quantum physics says that elementary particles exist as invisible waves extending in all directions until an observer looks at them. Then and only then does the particle assume a place in time and space. When he was out walking with Bohr, who was trying to convince him that quantum theory matched reality, Einstein pointed to the moon and said, "Do you really think the moon isn't there if you aren't looking at it?"

As the history of science turned out, Einstein was on the losing side of the argument. As Bruce Rosenblum and Fred Kuttner explain in their insightful book, *Quantum Enigma*, "Physicists in 1923 [were] finally forced to accept a wave-particle duality: A photon, an electron, an atom, a molecule—in principle any object—can be either compact or widely spread-out. You can choose which of these contradictory features to demonstrate." This sounds technical, but the punch line isn't: "The physical reality of an object depends on how you choose to look at it. Physics had encountered consciousness but did not realize it."

The fact that the physical world isn't a given has been validated over and over. This fact has huge importance for your brain. Everything that makes the moon real to you—its white radiance, the shadows that play across its surface, its waxing and waning, its orbit around the Earth—happens via your brain. Every aspect of reality is born "in here" as an experience. Even science, objective as it tries to be, is an activity taking place in consciousness.

On an everyday basis, physicists ignore their earthshaking discoveries about the quantum realm. They drive cars, not clouds of energy, to work. Once their cars are parked, they stay put. They don't fly off into invisible waves. Likewise, a brain surgeon cutting into gray matter accepts that the brain under his scalpel is solid and firmly placed in time and space. So when we want to go deeper than the brain, we must journey to an invisible realm where the five senses are left behind. We would have no urgent reason to take the journey if reality was a given, but it's not, with a vengeance. We will heed the words of Sir John Eccles, a famous British neurologist who declared, "I want you to realize that there exists no color in the natural world, and no sound—nothing of this kind; no textures, no patterns, no beauty, no scent."

You might feel a kind of existential queasiness trying to imagine what *is* out there if not color, sound, and texture. Reducing colors to vibrations of light won't solve anything. Vibrations measure light waves, but they say nothing about the experience of seeing color. Measurements are reductions of experience, not a substitute for them. Science rejects the subjective world, where experiences occur, because it is fickle, changeable, and not measurable. If person A loves Picasso's paintings and person B hates them, those are two opposite experiences, but you can't assign a number value to them. Brain scans don't help, either, since the same areas in the visual cortex will be active.

Where does solid ground lie when everything shifts and changes? You can't live in a world that rests on slippery illusion. As we see it, the way out is to realize that science is fooled by its own reality illusion. By rejecting subjective experiences like love, beauty, and truth, and substituting objective data—facts that are supposed to be more reliable—science gives the impression that vibrations are the same as colors and that electrons bouncing off electrons in the brain are the equivalent of thinking. Neither is true. The reality illusion needs to be dissolved, and that can only be done by discarding some outworn assumptions

SWEEPING AWAY THE REALITY ILLUSION
OLD BELIEFS THAT NEED TO GO

• The belief that the brain creates consciousness. In reality, it's the other way around.

- The belief that the material world is solid and reliable. In reality, the physical world is ever-shifting and elusive.
- The belief that sight, sound, touch, taste, and smell match the world "out there." In reality, all sensations are produced in consciousness.
- The belief that the physical world is the same for all living things. In reality, the physical world we experience only mirrors our human nervous system.
- The belief that science deals in empirical facts. In reality, science organizes and gives mathematical expression to experiences in consciousness.
- The belief that life should be lived by common sense and reason. In reality, we should feel our way through life utilizing as much awareness as we can.

Now we are diving into the thorny arguments we promised you, but the reassuring physical world vanished over a hundred years ago when quantum reality took over. It baffles physicists, as it does everyone else, to see the moon and stars vanish. With a mournful sense of finality, like a priest presiding over a casket, the French theoretical physicist Bernard d'Espagnat intones, "The doctrine that the world is made up of objects whose existence is independent of human consciousness turns out to be in conflict with quantum mechanics and with facts established by experiment."

Why should you and I care personally about this? Once each of us makes peace with reality instead of illusion, so many more possibilities exist—infinite possibilities, in fact. There is no need to be mournful. The mind has always amazed itself. Now it has a chance to fulfill itself.

Qualia

Human beings are incredibly fortunate that our brains can adapt to anything we envision. In the terminology of neuroscience, all the colors, sounds, and textures that we experience are lumped together under the term *qualia,* which is Latin for "qualities." Colors are qualia, and so are smells. The feeling of love is qualia; for that matter, so is the feeling of just being alive. We are like trembling antennas turning billions of bits of raw data into the bustling, noisy, colorful world—a world composed of qualities. So every experience is a qualia experience. The word is so bland that you'd never suspect qualia could become a baffling mystery, but it has.

It's inescapable, according to quantum physics, that physical objects possess no fixed attributes. Rocks aren't hard; water isn't wet; light isn't bright. These are all qualia created in your consciousness, using the brain as a processing facility. The fact that a physicist drives a car to work instead of a cloud of energy doesn't mean the invisible cloud of energy can be dismissed. It occupies the quantum level, where time is born, and space, and everything that fills space. You cannot experience time unless your brain interfaces the quantum world. You cannot experience space, either, or anything that exists in space.

Your brain is a quantum device, and somewhere below the level of the five senses, you are a creative force. Time is your responsibility. Space needs you. It doesn't need you to exist; it needs you to exist *in your reality.* If that sounds confusing, here's a telling example. A sixth sense exists that most people overlook, the sense of where your body is, including its shape and the position of your arms and legs. This sense is called proprioception. Knowing where your body is involves receptors in your muscles as well as sensory neurons in the inner ear, joined to your sense of balance, which is centered in the cerebellum. It's a complex circuitry, and when it breaks down, people have the eerie feeling of being disembodied. They do not know, for example, if they are holding their right arm up in the air, straight out, or down by their sides. Such cases are very rare, and fascinating. One way for someone who lacks proprioception to feel that they have a body is to ride in a convertible with the top down.

The wind rushing around them, as detected by working receptors in the skin, substitutes for the lost sixth sense.

In other words, the sensation of being wrapped by the wind gives these people a place in space. Since that sensation occurs in the brain, space needs the brain in order to exist. If a neutrino had a nervous system, it wouldn't recognize our sense of space, because a neutrino is a subatomic particle that can travel through the Earth without slowing down—to it, Earth is empty space. By the same logic, time also needs the brain, as is easily shown when you go to sleep and time stops. It doesn't stop in the sense that all clocks wait for you to wake up in the morning. Time stops *for you*.

Once you take away all the qualities that the brain is processing, the world "out there" has no physical properties remaining. As the eminent German physicist Werner Heisenberg stated, "The atoms or elementary particles themselves are not real; they form a world of potentialities or possibilities rather than one of things or facts." What's left when atoms and molecules vanish is the creator of those "potentialities or possibilities." Who is the elusive, invisible creator? Consciousness.

Finding out that you are a creator is an exciting prospect. We need to know more. A specialist in perception, cognitive scientist Donald D. Hoffman at the University of California at Irvine, coined a useful term: "conscious agent." A conscious agent perceives reality through a specific type of nervous system. It doesn't have to be a human nervous system. Other species are conscious agents too. Their brains interface with time and space, although not the way ours does. A tree sloth in South America might move a few yards in a day, a pace we—but not it—would consider excruciatingly slow. Time feels normal to a sloth, just as it does to a hummingbird that is beating its wings eighty times per second.

Here we are challenging one of the core beliefs that keeps the reality illusion going strong, the belief that the objective world is the same for every living thing. Using somewhat technical language,

Hoffman mounts a startling attack on this belief: "Perceptual experiences do not match or approximate the properties of the objective world but instead provide a species-specific user interface to that world." If you've stayed with the logic so far, this sentence will be clear to you, except for the phrase "user interface," which is adopted from computers.

Imagine the universe as an experience rather than a thing. You can experience what looks like a vast part of the cosmos by staring at the banquet of stars spread out on a clear summer night, but those stars aren't even a billion-billionth of the whole. The universe cannot be grasped without an infinite nervous system. Because of its quadrillion synapses, the human brain gives infinity a run for its money. Still, you would never be able to see, hear, or touch anything if you had to be in touch with your synapses—simply opening your eyes requires thousands of synched signals. So Nature devised a shortcut, which looks a good deal like the shortcuts you use every day on your computer. With a computer, if you want to delete a sentence, you simply push the delete key. You don't have to go into the machine's innards or fiddle with its programming. You don't have to rearrange thousands of zeroes and ones in digital code. One touch suffices—that's how a user interface works. In the same way, when you create qualia, like the sweetness of sugar or the brilliance of an emerald, you don't have to go inside your brain or fiddle with its programming. You open your eyes, you see light—and bingo, the whole world is suddenly there.

Arguing in this way, Hoffman has made himself a brave target. Arrayed against him is the entire camp of scientists who declare that the brain creates consciousness. Hoffman turns it around and says that consciousness creates the brain. Neither camp has an easy job proving its case. The "brain first" camp must show how atoms and molecules learned to think. The "consciousness first" camp must show how mind creates atoms and molecules. The cleverness in Hoffman's position—and we thank him heartily for his careful reasoning—is that he doesn't have to commit himself to explaining ultimate reality, a problem that defies reason. Is God the ultimate reality? Did your universe spring from an infinite number of multiverses? Did Plato hit upon the right idea thousands of years ago when he said that material existence is based on invisible forms?

Too many theories clash, but if you stick with the user interface—Nature's shortcut—locating ultimate reality doesn't matter. Physicists can drive cars to work and still know that cars are actually invisible energy clouds. What matters is that a nervous system creates a picture to live by. Just as time and space need to be real only *for you*, so does everything else. Religionists and atheists can sit down to tea together without fighting. The argument over ultimate reality will be unsettled for a long, long time. Meanwhile each of us will continue to create our personal reality—and hopefully get better at it.

Chasing the Light

If you can accept that you are a conscious agent, we're with you. But there is a nagging question left to settle. What is a conscious agent actually doing? In the Book of Genesis, God said, "Let there be light," and there was light. You are joining in that creative act this very minute, only you don't need words. (Neither did God, probably.) Somewhere in silence the most basic building block in creation, light, comes into reality the instant you open your eyes. If you are making light real for you, how do you do it?

Let's backtrack 13.8 billion years. At the instant of the Big Bang, the cosmos erupted from a void. Physics accepts that every particle in the universe is winking in and out of the void at a rapid rate, thousands of times per second. There are various terms for the void: the vacuum state, the pre-created state of the universe, the field of probability waves. The essential concept is the same, however. Far more real than the physical universe is the field of infinite potential from which it springs, here and now. Genesis never stopped at the level of the quantum field—all events past, present, and future are embedded there. So are all the things we can imagine or conceive of. That's why it would take an infinite nervous system to actually perceive "real" reality.

Instead, we make brain pictures that we call reality, even though these pictures are very limited. The only world that exists for human beings mirrors the evolution of the human nervous system. Brain pictures evolve. The way a physicist looks at fire isn't the same as the way Cro-Magnon man looked at fire—and probably worshiped it. Suddenly we see why the lower brain wasn't dropped or bypassed as the brain went on to higher things. All previous versions of a nervous system—going back to the most primitive sensory responses of one-celled organisms swimming toward the sunlight in a pond—are wrapped up and incorporated into the brain you have today. Thanks to your neocortex, you can enjoy Bach, which would be scrambled noise to a chimpanzee—but if an insane listener shoots the harpsichordist, you will react with all the primitive power of fight or flight in the reptilian brain.

The human brain didn't evolve on its own; it was following a picture of the world that existed in consciousness. The user interface kept improving to keep up with what the user wanted to do. At this moment, you own the latest version of the interface because you are participating in the latest "world picture" that humans have evolved to.

Whew.

According to Hoffman's theory, which he calls conscious realism, "the objective world consists of conscious agents and their perceptual experience." Goodbye to everything "out there"; hello to everything "in here." In fact, the two are merged at their source. Consciousness has no trouble weaving together both halves of reality. Now we come to the moment when you have to fasten your seat belt. There is actually no world "in here" or "out there." There is only the experience of qualia. Atoms and molecules aren't things; they are mathematical descriptions of experience. Space and time are also only descriptions of experience. Your brain isn't responsible for any of it, because your brain too is just an experience that your mind is having.

This is a huge leap, but it gives us untold power. Literally untold, for our parents and the society around us didn't tell us who we really are. We are the source of qualia. We are the caretakers of consciousness who do not need to bend before the forces of Nature. In our hands we hold the key to make Nature bend to us. Despite our limited minds, we are commanding "Let there be light" just as God does in his infinite mind. And yet this knowledge doesn't actually unlock the power. If you stand on a railroad track before an oncoming train, muttering "I created this reality," your mind won't prevent the enormous mass of a diesel engine from colliding with the small mass of your body, leading to unfortunate, messy results.

The ancient sages of India were not deterred by diesel trains (if they had existed back then), declaring that the world is only a dream. If a train hit you in a dream, you might feel all the sensations of being hit in real life, but you can wake up from a dream. There's the difference. Waking up from a dream seems easy and natural to us. Waking up from physical reality seems all but impossible, and while we remain in this representational world that we call physical reality, its rules of behavior follow Newton's Laws of Motion. Is that final?

One time a sorcerer took the hand of his apprentice and told him to hold on tight. "See that tree over there?" the sorcerer said, and suddenly he jumped over the top of the tree, taking the apprentice with him. When they landed on the ground, the apprentice went into severe distress. He felt dizzy and confused; his stomach turned over, and he started to retch. The sorcerer looked on calmly. This was just the mind's reaction to being shown its self-delusion. The mind cannot believe that it is possible to leap over a tree in real life as easily as in a dream.

We know that dreams all happen in our heads; we overlook that the waking state is also happening in our heads. But once the mind is shown its mistake, a new reality dawns. You may recognize this incident from the writings of Carlos Castaneda and his famous teacher, the Yaqui sorcerer Don Juan. Of course, any sensible person knows that those books are fiction.

Yet waking up from the dream is the key to enlightenment, as we saw in the last chapter. It is the basis of Vedanta, the oldest spiritual tradition in India, which spread its influence throughout Asia. A key concept in Vedanta is *Pragya paradha*, translated as "the mistake of the intellect" or some variant. The mistake comes down to forgetting who you are. Seeing ourselves as separate, isolated beings, we surrender to the look of the world, accepting that mindless natural forces control us. We are not taking a stand about jumping over trees or standing on railroad tracks. The waking state has its rules and limitations. The whole qualia argument attempts to return to the natural, basic act of perception, showing that reality isn't a given. We perceive what our nervous system has evolved to perceive.

To turn theory into practice, let's take this new viewpoint and see how it might change your life.

POWERING UP THE INTERFACE

There is no knowable reality without consciousness. You can create any quality (qualia) you want.

Everyone is creating qualia already. The secret is to become better at it.

To become better, you must get closer to the creative source.

The creative source is a field of infinite possibilities.

That field is everywhere, including inside your own awareness.

Capture the source of pure awareness, and you will have all possibilities within your reach.

This sequence represents knowledge that is thousands of years old, coming from sages who were Einsteins of consciousness. Once you return to your source, which is pure consciousness, you regain control over qualia. If you are receiving negative messages about your life—which can be negative thoughts "in here" or negative events "out there," these are all qualia. Which means that they can be changed if you change your consciousness.

Regaining control over qualia is the key to reshaping the brain and your personal reality at the same time. Sages and seers in the Eastern tradition would greet this argument with a smile and a shrugged "of course." In a materialistic age, it makes jaws drop.

By now, some readers may be crying foul play. Here they are reading a brain book, and suddenly the brain has vanished! It has been replaced by all-pervading consciousness. Skeptics will have none of it (believe us, we've butted heads with them). They won't budge from a stubborn insistence that consciousness *is* the brain. But Hoffman doesn't back down. He picks up this book's basic premise, that you are the user of your brain, not the other way around, and takes it to the limit: "Consciousness creates brain activity and the material objects of the world." In other words, we aren't machines that learned to think; we are thoughts that learned how to make machines. Once you accept this, the entire reality illusion explodes.

Consciousness Outside the Brain

Having gotten this far, which side do you think is right? If you believe that your brain is the creator of consciousness, then the materialists can win every argument. And not just materialists—also atheists, who believe that the mind dies when the brain dies. We can include too those people who have no ax to grind against God but simply accept that rocks are hard, water wet, and so on, all the commonsense experiences that hold the everyday world together. But the truth will out, and if it's true that consciousness comes first and the brain second, there has to be evidence for it.

Let's turn, then, to experimental proof. As early as the 1960s, pioneering researchers T. D. Duane and T. Behrendt demonstrated that brain-wave patterns of two distant individuals can sync with each other. The experiment involved the EEGs of identical twins. (This was decades before modern brain-imaging techniques like MRIs.)

In order to test anecdotal reports that twins share the same feelings and physical sensations, even when they are far apart, the researchers altered the EEG pattern of one twin and observed the effect on the other. In two of fifteen pairs of twins, when one twin closed his eyes, it produced an immediate alpha rhythm not only in his own brain but also in the brain of the other twin, even though he kept his eyes open and was sitting in a lighted room.

Were they participating in a shared mind, which is what some identical twins feel (although not all)? Striking anecdotes reinforce this finding. In his probing book, *The One Mind*, Dr. Larry Dossey presents the Duane-Behrendt study and relates a story in support:

> One case involved the identical twins Ross and Norris McWhirter, who were well known in Britain as co-editors of the *Guinness Book of Records*. On November 27, 1975, Ross was fatally shot in the head and chest by two gunmen on the doorstep of his north London home. According to an individual who was with his twin brother Norris, Norris reacted in a dramatic way at the time of the shooting, almost as if he too had been shot "by an invisible bullet."

Related studies prove that one mind can connect with another, as indicated by brain-wave correlations. (Rudy himself is a fraternal twin with his sister Anne. To his amazement, when he has a sudden urge to call her, he finds that she is feeling physically or mentally under the weather—somehow he senses that something is wrong.) Not just twins—nursing mothers are in sync with their babies and healers with their patients. In the framework of materialism, the existence of healers is scoffed at, but Dossey cites a pioneering study of native Hawaiian healers led by the late Dr. Jeanne Achterberg, a physiologist of the mind-body connection who was fascinated by anecdotes that native healers often did their work from a distance.

In 2005, after a two-year search, Achterberg and her colleagues gathered eleven Hawaiian healers. Each had pursued their native healing tradition for an average of twenty-three years. The healers were asked to select a person with whom they had successfully worked in the past and with whom they felt an empathic connection. This person would be the recipient of healing in a controlled setting. The healers described their methods in a variety of ways—as praying, sending energy or good intentions, or simply thinking and wishing the highest good for their subjects. Achterberg simply called these efforts distant intentionality (DI).

Each recipient was isolated from the healer while undergoing an fMRI of their brain activity. The healers were asked to randomly send DI at two-minute intervals; the recipients could not have anticipated when the DI was being sent. But their brains did. Significant differences were found between the experimental (send) periods and control (no-send) periods in ten out of eleven cases. For the send periods, specific areas within the subjects' brains "lit up" on the fMRI scan, indicating increased metabolic activity. This did not occur during the no-send periods. Dossey writes, "The areas of the brain that were activated included the anterior and middle cingulate areas, precuneus, and frontal areas. There was less than approximately one chance in 10,000 that these results could be explained by chance."

Buddhism and other Eastern spiritual traditions view compassion as a universal condition, shared by the human mind as a whole. This study offers support by showing that compassion being sent by one person can exert measurable physical effects on another person at a distance. Empathic bonds are real. They can cross the space that seems to separate "me" from "you." This connection isn't physical; it's invisible and extends outside the brain.

Thinking this way doesn't come naturally anymore, although over 80 percent of people, if asked whether God exists, still say yes. God must have a mind if he (or she) exists, and it would be impossible to argue that God's mind was created inside the human brain. It makes people uncomfortable to shake their worldview, however, even when the evidence—from physics, brain studies, and the experience of sages and seers for thousands of years—offers a new reality. Since a new reality would benefit every one of us, let's go into the lion's den and show why consciousness could not *possibly* be created by the brain.

In January 2010 Ray Tallis, who is described as a polymath, atheist, and physician, mounted a pointed challenge to "the brain comes first" position. His article in the journal *New Scientist* was titled "Why You Won't Find Consciousness in the Brain." As a "neuroskeptic," Tallis attacks the most basic evidence that makes scientists believe that the brain creates consciousness: those by-now-familiar fMRI scans that show regions of the brain lighting up in correlation with mental activity. At this point the reader already knows a good deal about them. Tallis repeats some of the points we've been making.

One of the first things a scientist is taught is that a correlation isn't a cause. Radios light up when music plays, but they don't create music. Likewise, one could argue that brain activity doesn't create thoughts, even though we now can see which areas are lighting up.

Neural networks map out and mediate electrical activity. They aren't actually thinking.

Electrical activity isn't the same as having an experience, which is what happens in consciousness.

Warming to his subject, Tallis offers other very telling challenges, such as the following. Science hasn't come close to explaining how it is that we can see the world as a whole but can also pick out details if we want to. Tallis calls this "merging without mushing." You can look into a crowd and see it as a sea of faces, for example, but you can also pick out a face you recognize. "My sensory field is a many-layered whole that also maintains its multiplicity," writes Tallis. No one can describe how a neuron has this ability, because it doesn't.

Asking the brain to "store" memory is impossible, Tallis contends. Chemical and electrical reactions happen only in the present. A synapse fires now, with nothing left over from the previous minute, much less the distant past. After the firing is over, the chemical signals that cross the synapse reset to their default position. The brain can strengthen certain synapses while weakening others through a process called *long-term potentation*. This is how certain memories become hardwired, while others do not. The question is whether the brain is capable of remembering what it did in the past, or is it actually consciousness that does this. Salt can dissolve only at the moment when you stir it into a glass of water. It can't store a memory of dissolving in water in 1989.

Tallis notes that there are even more basic issues, such as the self—no brain location has been found for *I*, the person who is having an experience. You simply know that you exist. Nothing lights up in your brain; no calories are expended to keep your sense of self going. For all intents and purposes, if the self had to be proven scientifically, a skeptic could examine brain scans and prove that there is no *I*, except that obviously there is, brain scans or not. *I* is actually operating the whole brain. It is creating pictures of the world without jumping into the picture, just as a painter creates paintings without jumping into them. To say that the brain creates the self is like saying that paintings create their painters. It doesn't hold up.

Then there is the initiation of action. If the brain is a biological machine, as materialists agree (a famous phrase from an expert in artificial intelligence dubs the brain "a computer made of meat"), how does the machine come up with new, unexpected choices? The most powerful computer in the world doesn't say "I want a day off" or "Let's talk about something else." It has no choice but to follow its programming.

So how can a machine made of neurons change its mind, have a spontaneous impulse, refuse to behave reasonably, and do all the other tricky things we do on a whim? It can't. This leads to free will, which strict determinism must deny. We all feel free in a Chinese restaurant to pick one dish from column A and another from column B. If every reaction in the brain is predetermined by the laws of chemistry and physics—as brain scientists insist—then the food you will order a week from now, or ten years from now, must be beyond your control. Which is absurd. Are we prisoners of the laws of physics or prisoners of our own blind assumptions?

Tallis's reasoning is devastating, but it was easy to dismiss as philosophy, not science. (To echo a familiar phrase that crops up when a scientist's thinking wanders beyond the accepted borders, "Shut up and calculate.") Neuroscience can chug along without answering such challenges, using the defense that each riddle will be solved sometime in the future. No doubt many will (and Rudy is part of the effort). Unless the link is made to show how atoms and molecules learned to think, however, the scientific picture of reality will be fatally flawed.

We feel that the burden of proof has been met. The thorn patch has been crossed. What's left is to show how you can master the qualia in your life. Negative signals can be turned into positive ones. More important, you can embrace the next step in your own evolution.

SUPER BRAIN SOLUTIONS

WELL-BEING

Happiness is hard to attain and even harder to explain. But if you want to experience a state of well-being—defined as overall happiness and good health—the brain must send positive messages instead of negative ones. What does *positive* mean? It has to be more than a surge of pleasurable impulses when you have a nice experience. Cells need positive messages in order to survive. So let's define positive as a qualia state. If the quality of your life is constantly being enhanced, its sights, sounds, tastes, and textures will always be shifting, but instead of being a chaotic mixture, there will be a lifelong trend in the direction of well-being.

The ingredients of well-being are yours to create and maintain. The controls exist "in here." Take two people who have identical work, incomes, houses, social background, and education. Included in these things are years of experience. But each person processes their experience differently. At fifty, Mr. A feels tired, restless, a bit bored, and cynical. His enthusiasm for life is starting to wear out. He wonders if anything new will revive his spirits. Mr. B, on the other hand, feels young, engaged, and vital. He sees new challenges around the corner. If you asked, he would say that fifty is the best time in his life.

Clearly the two men have a markedly different level of well-being. What made the difference? In terms of the brain, all experience must be processed through chemical pathways, much as the raw energy in food is metabolized. Chemical processing looks the same in every healthy cell. If you could measure metabolism by watching every molecule of water, glucose, salt, and so on passing through the cell membrane, the quantities being used would be so close that any two people should be processing experiences the same way. But they aren't. The metabolism of experience—which is what your brain is doing—depends on the quality of life, not the quantity. That's why we have leaned so heavily on qualia.

Well-being is a state in which experience has the following overall quality as it is metabolized in the brain:

You subtly feel that everything is okay.
You accept that you are okay.
There's a freshness to new experiences.
You enjoy the flavor of your experiences.
You spend each day emphasizing the positive possibilities
    and countering the negative implications.

These are qualities your brain registers, but it doesn't create them, for the simple reason that our brain can't have experiences. Only you can, and therefore you add the qualities of life, whether they are positive or negative.

Eavesdropping on your moods, beliefs, wishes, hopes, and expectations, brain cells are able to detect the quality of life. Neuroscience cannot measure this ongoing process, because it is concerned with the data measured by chemical and electrical activity. Minute as the changes are, over time the quality of life leaves biological markers. Everyone's brain displays markers of subjective states like depression, loneliness, anxiety, hostility, and general stress. Yet ironically, positive states tend to look rather flat and normal on a brain scan. Only in exceptional cases, such as the brains of long-term meditators, can you view unusual changes. For both sides of the coin, enjoying a low or high level of well-being can be traced back to how experience gets metabolized day by day, moment by moment, second by second.

Metabolizing Experience

The upshot is that you can improve your well-being by attending to subtle subjective cues. How often have you heard someone say,

"This doesn't pass the smell test"? Why are psychologists now giving weight to immediate reactions as being more reliable than long, rational consideration? This shouldn't be a new discovery. We've been living with human nature a long time. But the subtle instincts that enable you to feel your way through life are easily censored. Your mind leaps in with all kinds of secondary responses that are not good for you. These include

Denial—I don't want to feel this.

Repression—I keep my true feelings out of sight, and now I hardly know where they are.

Censorship—I only let good feelings register. The bad ones must stay away.

Guilt and shame—These are so painful that I must push them away as quickly as possible.

Victimization—I feel bad, but I don't deserve any better.

We are all familiar with these psychological mechanisms. Taken to extremes, they send millions of people into therapy. Unfortunately, you can feel basically all right and still be damaging your well-being by tiny degrees. A life of white lies, avoidance, judgment, self-abnegation, and petty illusions sounds harmless enough, but like Chinese water torture, negativity works by droplets. If you see someone leading a bitter or empty existence, it's usually not some huge melodramatic event that made them that way. Well-being was slowly worn down.

Well-being depends on many things going right in your nervous system. You can't attend to them one by one; infinitely too many processes are happening in the blink of an eye. Despite this complexity, you can begin to pay attention to subtle cues. In the Indian tradition, there are three classes of subtle cues wrapped inside every experience.

Tattva: the qualities or aspects of the experience
Rasa: the flavor of the experience
Bhava: the mood or emotional tone of the experience

Let's see how these are packaged into every experience. Imagine that you are on vacation sitting at the beach. The *qualities* of the experience would be your sense of the warm sun, the sound of the surf, and the swaying palm trees—the composite sensation of being on a beach. The *flavor* of the experience is subtler. In this case, let's say it's a sweet, relaxing experience that makes your body feel as if it is flowing into the whole beach scene. Finally, the *mood* of the experience isn't determined by either of the above. If you are lying on the beach feeling lonely or having a fight with your spouse, a beach isn't the same as it is to someone who is on a blissful honeymoon or is simply soaking up a nice tropical day.

Well-being is created at the subtle level. Therefore, as raw data stream into your brain through the five senses, what turns them into something nourishing or something toxic depends on the quality, flavor, and emotional mood that you add. We aren't discounting the brain, since of course it is a vital part of the mind-body feedback loop. There are neural networks that predispose you to have a positive or negative reaction automatically. But neural networks are secondary. What is primary is the person who is interpreting every experience as it is happening.

Subtle but Important

Instead of thinking all the time about how your life should be going, try a different tack. Learn to rely on the most holistic power you have, which is feeling. Feeling comprises the subtle underpinning of everything. Let's take one example from *rasa*, the flavor of life. According to Ayurveda, the traditional knowledge of medicine and overall health, there are six tastes: sweet, sour, bitter, and salty (the usual four), along with pungent (i.e., the spiciness of chilies and hotness of onion and garlic) and astringent (the mouth-puckering taste of tea, green apples, and grape skins).

Ayurveda takes the concept of *rasa* beyond what the tongue tastes. There is something subtler and more pervasive about the flavor of life. You can see this with the words we use in English.

We say *bitter* greens but also a bitter dispute, a bitter divorce, a bitter memory, and bitter relationships.

We say *sour* lemon but also sour grapes (meaning envy), sour mood, a sour note in music, and deals gone sour.

Each of the six *rasas* seems to have a root experience—they are like a family of flavors that pervade your life. In Ayurveda, if sweetness goes out of balance, the result can be obesity and putting on fat, but there is also a mental link to lethargy and anxiety. This is too vast a subject to cover here (and too alien to Western medicine for easy explanation). But anyone can look at the flavor of their lives and assess the difference, for example, between a sweet existence and a sour one.

In terms of *tattva*, or qualities, a personal connection goes beyond the five senses. Red, for instance, can be measured as a certain wavelength in the visible spectrum of light, but red is also hot, angry, passionate, bloody, and a warning. Green is more than a wavelength along the spectrum from red. Green is cool, soothing, fresh, and reminiscent of spring. What is crucial is to realize that these human qualities are more basic to existence than the measurable ones that science reduces to data. If you faint at the sight of red or feel buoyant at the first greening of spring, it's not wavelengths of light you are responding to but a complex of qualities, flavors, and emotions that combine to create an experience.

Now, what is the best approach to this wild complexity, which is far too intricate to handle one bit at a time? You can feel your way to well-being by increasing the life-enhancing ingredients that in Sanskrit are called *sattva*, usually translated as "purity." A *sattvic* life has a holistic effect as you begin to refine your sensations in all departments:

HOW TO FAVOR PURITY

Add to the sweetness in your life, and decrease whatever feels sour and bitter.

Lower the stress between yourself and others—favor respect, dignity, tolerance, and congenial interactions.

Act out of love whenever you can. Be compassionate. (But don't force yourself into a rigid kind of positivity. Your role isn't to be a smile robot.)

Find a sense of reverence for Nature. Go out in Nature to appreciate its beauty.

Be calm within yourself. Don't add to the agitation around you.

Don't step on other people's subtle level of feeling. Be aware that every situation has a feeling and mood that you should respect.

Practice nonviolence. Do not kill or harm other life-forms.

Be of service. Let the world be as close to you as your family.

Tell the truth without harshness.

Do what you know to be right.

Seek the presence of the divine.

This is the outline of a simple, well-regulated life that avoids agitation and chaos. As a framework, it allows for a good deal of personal interpretation. You can decide what makes your life sweet, for example. In the Indian tradition, diet is central, and the *rasa*, or taste, of sweetness is preferred. A *sattvic* diet is supposed to give lightness to body and mind. It is primarily vegetarian, focused on fruits, milk, grains, nuts, and other sweet foods.

Life can't be sweet all the time. The original intent of the Vedic sages wasn't to call some *rasas* good and others bad. (Every *rasa*, including bitter and astringent, has its place in the metabolizing of experience.) The sages were intent on giving positive signals to the brain and receiving positive signals back. Since the brain is the creation of consciousness, *sattva* begins in your awareness. If you practice purity because you want to and it feels good, your brain will be able to operate with much higher self-regulation. The best self-regulation is automatic, but you need to instill it first. Then more and more you can leave things to your autonomic nervous system, confident that it will support the well-being of your cells, tissues, and organs. The result will be a happier, healthier, and more spiritually enriched life.

RUDY'S EPILOGUE

LOOKING AT ALZHEIMER'S WITH HOPE AND LIGHT

It's fascinating to connect mind and brain, but when the connection breaks down, there is terror. My professional life has been spent researching the dark side of the brain. In the Alzheimer's Genome Project, my laboratory continues to find the genes, more than one hundred so far, involved in the most common and devastating form of dementia. Writing this book gave me a chance to step back and consider the brain in a wider perspective. The more you know about the mind, the more your research into the brain starts to shape itself into new patterns and possibilities.

Cancer researchers feel tremendous urgency to find a cure, not unlike the immense time pressure that overshadows Alzheimer's. As life span increases, so will the number of cases. Already more than 5 million Americans and 38 million people worldwide suffer from the disease. By 2040 the United States is projected to harbor over 14 million patients and our planet, over 100 million cases if effective preventive therapies are not developed.

At present, genetic studies offer our best chance to one day eradicate Alzheimer's disease. By uncovering all the genes that influence risk for Alzheimer's, we will someday be able to reliably predict a person's risk for the disease early in life. For those deemed at highest risk, it will probably be necessary to test for presymptomatic detection starting at around thirty or forty years old. Brain changes occur decades before the first signs of memory loss begin to show up. In its cruel progression, Alzheimer's destroys the areas of the brain devoted to memory and learning. Our immediate hope would be to empower high-risk individuals with therapies that can stop the further progression of their illness before dementia strikes.

Once we have drugs that can do this, we hope to prevent Alzheimer's before any clinical symptoms of cognitive decline begin to manifest. This so-called "pharmacogenetic" strategy is based on "early prediction–early detection–early prevention." If the three can be linked, we will hopefully be able to stop Alzheimer's before it begins. It's a broad strategy that goes back all the way to preventing smallpox with a vaccine in infancy but has spread out to preventing lung cancer by not smoking. A similar strategy can be applied to other common age-related diseases such as heart disease, cancer, stroke, and diabetes.

Does Alzheimer's have a lifestyle component? That's a question we can't fully answer yet, but I want to prepare myself for the possibility. The next horizon is the mind. Any lifestyle change begins in the mind. You have to want to change in the first place, and then you must lead your brain to create new neural networks to support your decision. We already know that "use it or lose it" applies to the brain in general, especially when it comes to keeping memory sharp and intact over a lifetime. Teaming up with Deepak, we looked into the mind-body connection much more deeply. When we came up with our "ideal lifestyle for the brain," we weren't implying that it is Alzheimer's specific. We are also not saying that Alzheimer's strikes because the patient did not live their life the right way. Genetics and lifestyle combine to cause this disease in most cases. And some genetic factors are too difficult to overcome with healthy living.

Almost all of us inherit gene variations that either increase or decrease our risk for Alzheimer's. These gene variants combine with environmental factors to determine your lifetime risk for getting the disease. Major risk factors cover a gamut of possibilities, including depression, stroke, traumatic brain injury, obesity, high cholesterol, diabetes, and even loneliness.

The genes that influence one's risk for Alzheimer's fall into two categories: deterministic and susceptible. A small portion of the disease incidence (less than 5 percent) strikes under the age of sixty. It is most often due to mutations in one of the three genes that my colleagues and I discovered. These inherited mutations virtually guarantee onset of the disease in one's forties or fifties. Luckily, these gene mutations are pretty rare. In the vast majority of cases, Alzheimer's strikes after age sixty. In these cases, genes have been identified that carry variants influencing one's susceptibility. Such variants do not cause the disease with certainty, but when they are inherited, they confer either increased or decreased risk for the disease as a person ages.

The good news is that in most cases of Alzheimer's, one's lifestyle can potentially trump one's genetic predisposition for the disease. A similar genetic picture presents itself in most of the common age-related disorders, like heart disease, stroke, and diabetes. Can certain behaviors indicate a pattern in brain activity that could be treated early on? Some autism researchers are asking this question about infants who do not yet show signs of the disorder but who may be holding their heads up in a certain way that is a precursor to autism. One of the biggest advances in brain research has gone unnoticed by the general public. It is the turn from the synapse to the network. For decades neuroscience focused its main effort on how the single synapse, the communication junction between two neurons, actually works. The research work was grueling and meticulous. Imagine trying to stop lightning as its flashes through the sky, only on a scale millions of times smaller. The important breakthroughs, which came slowly, involved freezing brain tissue to extract the messenger molecules that became known as neurotransmitters. Studies of two of them, serotonin and dopamine, provided huge progress in treating disorders ranging from depression to Parkinson's disease.

But studying the synapse didn't carry us far enough. There are many different kinds of depression, for example, each with its own chemical signature. But broad-spectrum antidepressants were not effective in pinpointing each type, because in patient A the range of symptoms are likely not the same as in patient B, even though both fall into the constellation of sadness, helplessness, fatigue, sleep irregularity, lost appetite, and so on. Depression forms its own unique neural network from person to person.

That is why a systems approach arose, looking at bigger patterns of networks that extend far beyond the synapse. In your house, examining one fuse in the breaker box isn't all that different from looking at the entire wiring scheme. That's not so in your brain. Neural networks are alive, dynamic, and interrelated in such a way that a change in one piece of the wiring will reverberate throughout the whole nervous system.

As abstract as that sounds, the network approach opens a phenomenal number of doorways. We confront the brain as a fluid process, not a thing. Since thinking and feeling are also fluid processes, it's like watching two mirror universes. (The unconscious mind can even be seen as parallel to the "dark" matter and energy that mysteriously control events in the visible cosmos.) In this wide picture, your neurons behave in sync with everything that is happening to you, and even your genes participate. Far from sitting in frozen silence at the heart of each cell, your genes are switched on and off, changing their chemical output, according to all kinds of events in your life. Behavior shapes biology. Using that watchword, research has shown that positive lifestyle changes in diet, exercise, stress management, and meditation affect four hundred to five hundred genes—and probably many more.

What can you do to prevent or stave off the onset of Alzheimer's? Follow the lifestyle trend that is working elsewhere for so many disorders. For starters, exercise. A close colleague, Sam Sisodia, showed that in animal models (mice given the human Alzheimer's gene mutations), providing running wheels for exercise at night dramatically reduced brain pathology. Exercise actually promoted gene activity that lowered beta-amyloid levels in the brain. Epidemiology studies also have confirmed that moderate exercise (three times per week for one hour) can lower risk for Alzheimer's. One clinical trial indicated that sixty minutes of robust exercise twice a week was able to slow progression of the disease once it began.

The second key is diet. The rule of thumb is that if what you eat is good for your heart, it's good for your brain. A Mediterranean diet rich in virgin olive oil, as well as moderate amounts of red wine and even dark chocolate, has been associated with lower risk for Alzheimer's. An even simpler preventive is to eat less. In animal models, caloric restriction increases longevity and reduces brain pathology. (More recently, virgin coconut oil has been proposed to treat and prevent the disease. However, more data is needed to assess this claim.)

You are achieving the third means of prevention as you read this book. It is intellectual stimulation, which stimulates new synapses in the brain. Every new synapse you make strengthens those you already have. Like money in the bank, making more synapses means you won't be so easily depleted before getting Alzheimer's. While Alzheimer's affects people with the full spectrum of education from high school dropout to Ph.D., some studies suggest that higher levels of education can be protective. Perhaps more important than intellectual stimulation is social engagement. Being more socially interactive has been associated with lower risk, while loneliness has been documented as a risk factor for getting the disease.

It would be tremendous if Alzheimer's could make the same turnaround that has occurred in cancer. A decade ago cancer treatment was almost entirely focused on early detection followed by drugs, radiation, and surgery. The Centers for Disease Control estimated in 2012 that two-thirds of cancers are preventable through a proactive lifestyle, avoiding obesity, and not smoking. Other cancer centers raise that estimate to between 90 percent and 95 percent.

The signs of progress on all fronts—chemical, genetic, behavioral, and lifestyle—are encouraging. But they alone wouldn't have led me to write about super brain. In my field you can thrive by being a superb technician, carving out your scientific niche in the detailed analysis of very narrow aspects of a disease. You can make it pretty far in science by ceasing to speculate and obeying the dictum to "shut up and calculate." Hard science is proud of its status in society, but I have also witnessed firsthand that this pride can extend to arrogance when it comes to considering the contributions of meta-physics and philosophy to developing scientific theories. This broad dismissal of anything that cannot be measured and reduced to data strikes me as incredibly narrow-minded. How can it make sense to dismiss the mind, however invisible and elusive it may be, when science is entirely a mental project? The greatest scientific discoveries of the future often begin as pipe dreams of the past.

Super brain represents the efforts of two serious investigators, both coming from medicine, to see as far into the mind-brain connection as possible. It's a bold step for a "hard" brain researcher to take the position that "consciousness comes first," but the evolution of my thinking has gradually led me there—as it has led eminent figures like Wilder Penfield and Sir John Eccles before me. In my view, neuroscientists cannot afford to ignore the interface with consciousness, because by arguing that "the brain must come first," they could be guilty of protecting their turf rather than acting like real scientists in pursuit of the truth, wherever the trail may lead.

The truth about consciousness has to involve more than electrons bouncing off electrons inside the brain. I went into Alzheimer's research to solve a difficult physiological puzzle, but just as important was the stirring of compassion I felt, especially after I watched my own grandmother succumb to this terrible disease. When Alzheimer's strikes, the sufferers and their loved ones feel completely betrayed. Even the early stages are frightening. The earliest signs are "mild cognitive impairment," which sounds fairly harmless. Once it arrives, however, the human effect is hardly mild as a patient begins to have trouble keeping track of everyday activities and is no longer able to multitask. As words become harder to find, the patient will also have increasing difficulty speaking and writing.

Worse than this, however, is the sense of doom that sets in. There is no turning back once the process begins. Old memories vanish, and new ones cannot be formed. Eventually the sufferer becomes unaware that he has the disease, but by then the job of full-time caretaker has been passed on, mostly to the immediate family. It's estimated that 15 million unpaid caregivers are involved right now. This terrible thief of minds creates suffering all around it.

Compassion affects anyone who witnesses this epidemic first-hand, but we can strive to convert pity and doom into a different perspective. Why not take the reality of Alzheimer's as a spur to use our brains the best way possible in the decades before we grow old? Alzheimer's kills the dream that old age will be a fulfilling time of life. Before winning the victory of curing the disease, each of us can win another victory, by using our brains for fulfillment, even from childhood. That's the vision of super brain, the part of this book that means the most to me.

As a species, we should take time every day to be thankful for this amazing organ buzzing away in our heads. Your brain not only transmits the world to you but essentially creates that world. If you can master your brain, you can master your reality. Once the mind unleashes its profound power, the result will be greater awareness, a healthier body, a happier disposition, and unlimited personal growth. New discoveries will continue to astonish us regarding the brain's ability to regenerate and rewire its circuitry. That rewiring is physical, but it happens in response to mental intentions. We must never forget that the true seat of human existence is in the mind, to which the brain bows like the most devoted and intimate of servants.

DEEPAK'S EPILOGUE

BEYOND BOUNDARIES

The full impact of super brain probably won't be realized for decades. We started out asking you to create a new relationship to the brain, mastering its wondrous intricacy. The best user of the brain is also an inspiring leader. We hope you have come closer to fulfilling that role. If so, you are the wave of the future. You will be making the next leap in the human brain's evolution.

Neuroscience is still reveling in its golden age, infatuated with matching areas of brain activity to specific behaviors. That's been a productive project, but it is running into contradictions, as it must when you try to reduce the mind to a physical mechanism. Human beings are not puppets being operated by the brain. Neuroscientists can't make up their mind about that, however. The latest research on drug addiction, for example, has become very specific about the damage done to opiate receptor sites by cocaine, heroin, and methamphetamines. This damage is considered permanent, and it leads to greater cravings for bigger doses. At a certain point, every drug addict stops getting high and maintains his self-destructive habit simply to feel normal.

This picture offers strong evidence that addiction is a cruel example of the drug using the addict, instead of the other way around. Some experts, citing the research, claim that addiction is all but impossible to break; the toxic chemicals exert an iron grip. And yet people do get over addictions. They confront their ravaged brains and manage to impose their own will. "I can kick this thing" is a cry that often fails but sometimes succeeds. It's a cry from the mind, not from the brain. It expresses choice and free will. Because choice and free will are unpopular among neuroscientists, we've worked hard in this book to restore them.

Our second goal has been to make higher consciousness believable. I welcomed the chance to work with a brilliant researcher because it's clear that modern people are not going to accept enlightenment without facts to back it up. The facts are there, in abundance. The brain will follow wherever the mind leads, even to the domain of God. Of all the messages sent by the brain, the subtlest ones, which are all but silent, hint at the divine. Millions of people don't heed these messages because silence gets overlooked in the rush and noise of daily life. But the whole ethos of science also makes it difficult to believe that God—an invisible being who leaves no traces in the physical world—might be real.

We take for granted lots of things that would not seem real if you measured them by visible physical evidence, starting with music and mathematics and ending with love and compassion. After writing this book, I realized that God isn't a luxury or an add-on to everyday existence. Beyond organized religion, which many are abandoning, people need consciousness to have a source.

If it didn't, we'd be in the position of Lois Lane, in a funny moment from the first *Superman* movie from 1978. Lois has been thrown off the top of a skyscraper and is plummeting to earth. Clark Kent, seeing her fall, jumps into a phone booth to change into his Superman costume for the first time. He soars up and catches Lois, saying, "Don't worry, miss, I've got you." Lois's eyes widen with fright. "But who's got *you*?" she cries.

The same question pertains to consciousness. It needs something or someone to uphold it, and that someone is the infinite consciousness we traditionally call God. If there were no God, he'd have to be invented. Why? Consider the argument that we've described as "the brain comes first." If consciousness arose from chemical interactions in the brain, as this argument contends, there is no need for God.

Atoms and molecules can take care of the business of mind on their own.

But we have argued that it is impossible for the brain to create consciousness. No one has come close to showing the magical transformation that allows salt, glucose, potassium, and water to learn to think. Modern society finds it primitive that our remote ancestors worshipped the spirits that inhabited trees, mountains, idols, and totems—a practice known as animism. Our ancestors were ascribing mind to physical objects. But isn't neuroscience guilty of animism when it claims that the chemicals in the brain are thinking? The reverse is far more plausible. Consciousness—the invisible agency of the mind—created the brain and has been using it ever since the first living organisms began to sense the world. As consciousness evolved, it modified the brain to its purposes, because the brain is only the physical representation of mind.

Turning the tables on neuroscience in this way seems shocking at first. But it gives God a new lease on life (not that he was ever dead). For a moment, rid yourself of any mental picture you have of God. Instead, imagine a mind with the same qualities as yours. It can think and create. It enjoys new possibilities; it can love, and the main thing it loves is being alive. This is the mind of God. What makes such a mind controversial is that it isn't localized. It expands beyond all boundaries. It operates in all dimensions without regard for past, present, or future. Every spiritual tradition has conceived of just such a God. However, this conception has deteriorated over time. Now we call God a matter of faith rather than a fact of Nature.

The brain restores God to being a fact. Once the argument that "brain comes first" falls flat, the only thing left is mind that sustains itself, mind that has always existed and pervades the cosmos. If this seems too hard to swallow, think back to medieval navigators who had learned to use lodestones, naturally magnetized pieces of the mineral magnetite. Suspended from a string, a lodestone will point north, functioning as a primitive compass. If you told a medieval navigator that magnetism existed everywhere, not just in a single stone, would he have believed you?

Today we walk around assuming that each of us has a mind, holding on to a prized piece of consciousness the way sailors once held on to lodestones. But the truth is that we participate in one mind, which hasn't lost its infinite status by existing in the small packages of individual human beings.

We are so attached to our own thoughts and desires that we easily say "my mind." But consciousness could be a field like electromagnetism, extending throughout the universe. Electrical signals permeate the brain, but we don't say "my electricity," and it's dubious that we should say "my mind." The pioneering quantum physicist Erwin Schrödinger made a flat statement about this on several occasions. Here are three:

"To divide or multiply consciousness is something meaningless."

"In truth there is only one mind."

"Consciousness is a singular that has no plural."

If that sounds metaphysical, it helps to remind ourselves that there is only one space and one time in the cosmos, even though we chop them up into small slices for everyday convenience.

One day science will catch up with all of these issues. The encounter is inescapable because it has already happened. The rock has fallen into the pond, and no one knows how far the ripples are spreading. Max Planck, who is credited with starting the quantum revolution more than a century ago, said something wonderful and mysterious: "The universe knew that we were coming." The mind field is at least as old as the universe, whereas the human brain is a product of evolution. Where will it evolve next? No one knows, but I plump for a giant leap so that we accept two words from ancient Sanskrit: *Aham Brahmasmi*, "I am the universe." This looks like a leap backward in time, but the Vedic seers spoke from a higher level of awareness. The passage of time doesn't cause "Who am I?" to become an outdated question. It would be astonishing if the everyday modern person caught up with ancient wisdom, but why not?

The brains of the Buddha, Jesus, and the rishis, or enlightened sages of India, reached a level that has inspired us for centuries, but as a biological creation, their brains were no different from that of any healthy adult today. The Buddha's brain followed where his mind led, which is why all the great spiritual teachers declared that anyone could make the same journey that they did. It's only a matter of setting your foot on the path and paying attention to the subtle signals picked up by your brain. Since it is attuned to the quantum level, your brain can receive anything that creation has to offer. In that sense, the great saints, sages, and seers weren't more favored by God than you and I are; they were braver about following a trail of clues that led them to the very source of their awareness.

If enlightened sages had been conversant in science-speak, they might have said, "The universe is an undivided wholeness in flowing movement." Instead, that sentence comes from the far-seeing English physicist David Bohm. It's the equivalent of "You can never step in a river in the same place twice." Thus do mystical conundrums reemerge as scientific hypotheses.

I'm an optimist, and I hope to see the validation of consciousness reach full scientific acceptance in the coming decade. The barriers that keep us earthbound are of our own making. They include the barrier that divides the world "in here" from the world "out there." Another barrier isolates the human mind as a unique product in the universe, which is otherwise devoid of intelligence—or so the prevailing theories of cosmology assert. In pockets of speculative thinking, however, a growing number of cosmologists have found the courage to look in a different direction, toward a universe teeming with intelligence, creativity, and self-awareness. Such a universe would indeed know that we were coming.

This book has touched upon many difficult concepts. There is one, however, that all the others depend on: reality making is every person's task. There is no real look to the world, no anchor we can drop once and for all. Reality keeps evolving (thank goodness), and the biggest clue to this lies inside your brain. One reality after another is packed into it. The reality of the reptilian brain is still in there, but it has been incorporated through evolution into higher realities, each one matched by a new physical structure.

The brain mirrors the reality that each person is making at this very moment. Your mind is the rider; your brain is the horse. Anyone who has ridden horses knows that they can balk, fight the bridle, become frightened, stop to munch grass by the wayside, or bolt for home. The rider hangs on, yet most of the time he is in command. We all relate to our brains by hanging on during the episodes when hardwired imprints, impulses, drives, and habits are in control. No horse has ever bolted as wildly as a brain gone awry. The physical basis for drug addiction, schizophrenia, and many other disorders cannot be denied.

Most of the time, however, mind is in the saddle. Conscious control is ours and always has been. There is no limit to what we can inspire the brain to achieve. It would be ironic if anyone turned away from super brain for being too unbelievable, because if you could only see your untapped potential, you would realize that you already own a super brain.

ACKNOWLEDGMENTS

Deepak Chopra

This book required the support of people who have become like an extended family, one that is always helpful, cordial, and never fights at Thanksgiving. At the Chopra Center my life is handled by Carolyn, Felicia, and Tori far better than I could handle it on my own. The same care is given to my writing by Julia Pastore, Tina Constable, and Tara Gilbride. My loving thanks go to all of you, and to my family at home, constant as ever.

It took more than two decades for me to consider working with a collaborator, and now that this phase has begun, let me acknowledge that Rudy has been the best of collaborators, an eye-opening example of an accomplished scientist who has a spiritual vision of life's possibilities.

Rudolph E. Tanzi

My contributions to this book would not have been possible without the endless support, advice, and inspiration of my loving wife, Dora, and the love of our beautiful daughter, Lyla. Throughout my life, I have been very fortunate to have my family always emphasize the importance of love and maintaining balance in one's mental and spiritual development. Thanks also to Julia Pastore, Tina Constable, and Tara Gilbride for sharing our passion and vision in making this book possible.

And, finally, I would like to thank Deepak for being the perfect collaborator and for becoming a dear friend and brother while writing this book together. Deepak's unique and wonderful outlook on the spiritual and scientific sides of the world, along with his impeccable ability to express it, has made writing this book a true joy.

INDEX